Great Thinkers of India Series. 1

RAMANUJAN — THE MAN AND THE MATHEMATICIAN

Ramanujan
The Man and The Mathematician

S.R. Ranganathan

National Research Professor in Library Science
Formerly Assistant Professor of Mathematics
Presidency College, Madras

Published By:

Ess Ess Publications

For

Sarada Ranganathan Endowment for
Library Science
Bangalore

Published by:
Ess Ess Publications
4837/24, Ansari Road,
Darya Ganj,
New Delhi-110 002.
Tel.: 23260807, 41563444
Fax: 41563334
E-mail: info@essessreference.com
www.essessreference.com

For:
Sarada Ranganathan Endowment for Library Science
702, 'Upstairs', 42nd Cross, III Block,
Rajajinagar,
Bangalore-560 010.
Tel. : 080-23305109
E-mail: srels@dataone.in

Rs.375/-

First Edition - 1967
First Reprint - 1989
Second Reprint - 2009

ISBN: 978-81-7000-557-5

Cover Design by Patch Creative Unit

Printed at Salasar Imaging Systems

PRINTED IN INDIA

to the memory of

V RAMASWAMY AYYAR

the founder of the Indian Mathematical Society
the first in the chain of the discovery of
Ramanujan

CONTENTS

A   INTRODUCTION   9

B   RAMANUJAN: A PUZZLE   12

C   SUPER-ACTIVITY PERIOD 1: 1907 to 1911   19

D   ECONOMIC SUFFICIENCY   23

E   UNIVERSITY SETTING   29

F   RAMANUJAN'S LIFE ABROAD   36

G   SUPER-ACTIVITY PERIOD 2: 1914 to 1918   40

H   PRECIPITATION OF HONOURS   44

J   RAMANUJAN'S HEALTH FAILS   46

K   MEMORIAL   51

L   RAMANUJAN'S NOTEBOOKS   55

M   REMINISCENCES OF FRIENDS   61

N   RAMANUJAN, THE MAN   92

P   RAMANUJAN, THE MATHEMATICIAN   105

Q   WORKS BY AND ON RAMANUJAN   118

R   UP-TO-DATE BIBLIOGRAPHY   123

S   BIO-DATA   126

     Index   129

LIST OF PLATES

(These photographs appear between pp 104 - 105)

All the photographs listed below except 1 and 6 are by Mr C Seshachalam, (Curzon & Co., Madras), President, Madras Amateur Photographic Society, who retains the copyright and to whom the publishers are grateful for permission to use them.

1   SRINIVASA RAMANUJAN
2   ENTRANCE DOOR OF RAMANUJAN'S HOUSE IN TRIPLICANE
3   INSCRIPTION ON MEMORIAL TABLET OVER THE ENTRANCE
4   MRS RAMANUJAN
5   RAMANUJAN COMMEMORATION STAMP ON "FIRST-DAY" COVER
6   COPY OF RAMANUJAN'S LETTER TO THE REGISTRAR, UNIVERSITY OF MADRAS, FROM A NURSING HOME IN PUTNEY, LONDON

CHAPTER A

INTRODUCTION

A1 Meteoric Career

Like a meteor, Srinivasa Ramanujan appeared suddenly in the mathematical firmament, rushed through the short span of his life, consumed himself and disappeared with equal suddenness. His precocity in Mathematics was noted very early by his teachers at school and by the boys of his age group in his home-town, Kumbakonam. His unusual insight into abstruse problems in Mathematics, came to the notice of senior mathematicians in Madras when he was about 23.   After his full discovery by G H Hardy, he lived only about six years — five in Cambridge and one in Madras.   He died in 1920 when he was 32.

A2 First Biography

A couple of years after the death of Ramanujan, the publication of his *Collected papers* was contemplated.   In 1923, it was decided that a biography of Ramanujan should be given at the beginning of his *Collected papers*.   The University of Madras appointed a Committee to write the biography.   It consisted of E M MacPhail, the Vice-Chancellor, P V Seshu Ayyar, Professor of Applied Mathematics in the Presidency College and Secretary of the Indian Mathematical Society, and R Ramachandra Rao, Education Secretary of the Government of Madras and former President of the Society.   As a junior member of the staff of the Department of Mathematics of the Presidency College, it fell to my share to prepare a draft of the biography.   As approved by the Committee, it was published in the *Collected papers* in 1927.

A3 Inaccessibility of the Biography

Only a select few, even among mathematicians, could find interest in the mathematical papers.   Therefore, to all the others the biography was virtually inaccessible. However, a brief version of it was brought out in some Indian languages in a pamphlet form for the use of school children.   On account of the peculiar

9

situation in India, the educated adults would read only in English. Therefore, these pamphlets too did not reach them.

A4 Ignorance about Ramanujan

Since about 1950, I often came across university graduates who were either totally ignorant of Ramanujan and his career, or had heard little about him except his having been a mathematician. By about 1960, the number of university graduates who had not even heard of his name had increased considerably. In my class on the "Universe of Knowledge: Its Development and Structure", in the Documentation Research and Training Centre, Bangalore, in 1962, I had to deal with the history of diverse subject-fields — the important epochs and the outstanding contributors to knowledge. My method of teaching is a discussional one. The development of the subject was through questions and answers. The class in DRTC consists of persons with a post-graduate degree. That year the class was only seven strong. To my surprise, more than half of them did not know about Ramanujan. This made me watch this ignorance a little more extensively. I found a similar ignorance about our eminent scientists of the nineteenth century among most of the younger university men. But they all had a fairly good knowledge about the politicians. One of the reasons for this was, that there were readable biographies of many of our statesmen and politicians but hardly any of our scientists. I traced this lacuna to the failure on the part of the next earlier generation to produce readable biographies of our scientists. But it did not occur to me that I myself should do something in the matter.

A5 Seventy-fifth Birthday of Ramanujan

In December 1962, the Union Government announced the proposal to issue a postage stamp with Ramanujan's figure. The Madras University also decided to celebrate the Seventy-fifth Birthday of Ramanujan. The All India Radio, Madras, decided to put on the air a talk on Ramanujan. Somehow they came to know that I had drafted the biography of Ramanujan included in his *Collected papers*; and they invited me to give a talk. After this talk I received letters from a couple of the surviving friends of Ramanujan. In the course of correspondence they mentioned some others of their group. This made me take up the collection of additional information about Ramanujan. Writing to Ramanujan's friends led to a kind of chain action. I got seventeen responses including the one from Mrs Ramanujan. It took me nearly two years to get the reminiscences from all of them.

A6 Genesis of this Biography

Thereafter, I started writing the biography. While doing so, it struck me that similar biographies of the other great thinkers of India should also be written. This made me write to my Publishers, the Asia Publishing House, Bombay. For, its Founder, Mr P S Jayasinghe, is well known for his enterprise and the patriotic zeal with which he pursues it. I suggested that the Asia Publishing House should run a new series called, "Great Thinkers of India"; and offered my draft biography of Ramanujan to be published as the first volume of that series. It was accepted. This is the genesis of this book.

A7 Conspectus

The Contents Page forms a clear Conspectus of the book. This biography largely concerns itself with the human aspect. My poor competence to deal with the mathematical achievements of Ramanujan has itself saved me from the temptation to load the biography with technical matter. Chapter P is the only one which has anything of a technical flavour. It will be found that it consists practically of extracts from the writings of G H Hardy. But the Librarian in me could not be dissuaded from including in this biography, Chapters Q and R, which are of a bibliographical import.

A8 Acknowledgement

My acknowledgement is due to all those, including Mrs Ramanujan, who have given their reminiscences. It is also due to the persons, whom I had contacted in their lifetime and who had then given me information of various kinds about Ramanujan.

RAMANUJAN: A PUZZLE

B1   A Phenomenon in Psychology

It was an evening in August 1913. Three Professors — P V Seshu Ayyar (Mathematics, Presidency College), Edward B Ross (Mathematics, Madras Christian College), and George Pittendrig (English, Madras Christian College) — were sitting in a group in the house of Prof Alexander Moffat (Physics, Madras Christian College). They were surrounded by some students. I was one among them. The discussion among the Professors was turned on Ramanujan. Seshu Ayyar was telling the Professors some details of the story of the discovery of Ramanujan. They were trying to explain his profound attainment in Mathematics without the help of books or teachers. Pittendrig said, "We cannot explain it. Ramanujan is a phenomenon." In fact, Ramanujan became one of the chief topics of talk among the elite of Madras during 1913 and 1914.

B2   A Polish Mathematician Anticipated

In November 1913, I was the only student in the post-graduate honours class in Mathematics of the Madras Christian College. Ross was my only Professor. We used to meet nearly three hours a day. One morning he entered the class-room with his eyes glittering and his lips throbbing. He asked me, "Does Ramanujan know Polish?" I replied that it was not at all likely. The Professor said, "Even if he did, it will make no difference." I was puzzled by this laconic remark of the Professor. Then he pulled out from his pocket a university envelope stuffed with a bunch of sheets. He threw the sheets open before me and said, "This is the quarterly report of Ramanujan as a research student of the University. Look at this beautiful theorem. In the issue of a Polish periodical brought by today's mail, something of this kind appears. Surely, Ramanujan could not have divined what that Polish mathematician was thinking. What is more, Ramanujan's theorem is much

12

deeper. Ramanujan has certainly anticipated the Polish mathematician. He is extraordinary. Is he not?"

B21 TRANS-RATIONAL INFORMATION
Some of the information collected for this biography and the earlier one by me is of a trans-rational nature.

B22 INITIATION OF PRECOCITY
One of them is a dream of Ramanujan. According to this dream, Goddess Namagiri — the deity of Namakkal — wrote on his tongue. Thereafter his precocity developed suddenly. It has been stated by his mother that he was born after her parents had prayed to the Goddess to bless her with a son. There is another piece of information current in Ramanujan's family. His maternal grandmother was a great devotee of Goddess Namagiri. She would often go into a trance and speak as Goddess Namagiri. In one such trance, before the birth of Ramanujan, she is said to have uttered that, after her own death, the Goddess would speak through the son of her daughter.

B23 PERMISSION TO GO ABROAD
Another piece of trans-rational information is about a dream that Ramanujan's mother had. Under the pressure of the then prevailing social tradition in India, she did not like the idea of her son crossing the seas. However, in 1914 she saw, in her dream, her son sitting amidst white persons, with a great halo about him. This dream made her ultimately agree to Ramanujan sailing to England. According to the information collected from Mr N Subbanarayanan, the son of S Narayana Ayyar who played no small part in the discovery of Ramanujan, Narayana Ayyar, himself, and Ramanujan spent three consecutive days in Namakkal to invoke the direction of Goddess Namagiri in the matter. It is only after this that Ramanujan's mother had the above-mentioned dream.

B24 SEEING DEATH
The third piece of information of trans-rational nature concerns a tragic event. This was narrated to me by G V Narayanaswamy Ayyar, a good friend of mine. He was a teacher in the Hindu

High School (Triplicane(. He was a reputed astrologer. He was
an old student of Prof Seshu Ayyar. It was March 1920. An
elderly lady came to Narayanaswamy Ayyar with a note from the
Professor. After reading the note, Narayanaswamy Ayyar asked
the lady for the horoscope. To his surprise, she began to dictate
the horoscope from memory. He spent some time studying the
horoscope. Then he asked her, "What do you want to know?"
The old lady asked about the longevity of the person. After some
deep thinking, Narayanaswamy Ayyar told her, "There is an
unusual indication in the horoscope. He is likely to attain world-
wide reputation and die at the height of his reputation. And if
he were to live long enough, he should be an obscure man. I find
it difficult to choose between these two indications. Who is this
gentleman? What is his name? Can you tell me?" The old
lady mentioned the name of Ramanujan. The astrologer felt
shocked that he should have told her so much without circum-
spection and reserve. He soon recovered himself and began to
mitigate his statement in some oblique way. The old lady replied,
"Sir, you need not hide anything from me. I myself suspected
all this." This was a surprise to the astrologer. The lady not
only remembered the details of the horoscope without the aid of
even a scrap of paper, but could also read the results from it.
Then the astrologer said in effect, "I am sorry that I was so hasty.
I was really so much carried away by the indication of such an
extraordinary attainment and reputation. This made me forget
an important rule of conduct among astrologers. Please do not
carry what I said to any of the relatives of Ramanujan." The
old lady with tears replied, "I am the mother of Ramanujan, that
unfortunate genius." She then burst into irrepressible sobs.
After her recovery from this fit of sobs, the astrologer said,
"Perhaps the wife's horoscope may have a mitigating influence.
Can you bring her horoscope next time?" The mother again
dictated straightway the horoscope of Janaki, her daughter-in-
law. The astrologer tallied the two horoscopes, compared them
and searched for any mitigating influence. After about half an
hour's search, he said, "It is desirable that he does not live with
his wife for a few months." The mother said, "I had seen it myself
in the horoscope. The wife's horoscope aggravates the indications
of my son's horoscope. I have been, therefore, pleading with
14

my son to send her away for sometime to her parents' house. But, alas! he utterly refuses. He has been always an obedient son. But in this particular matter, his obstinacy is unbreakable."

And Ramanujan died on 26 April 1920.

B3 Common Allergy to Trans-rational Phenomenon

It was early morning one day in January 1929. A probationer of the Indian Civil Service, a former student of mine in the Presidency College (Madras), called on me at the University Library. Our conversation soon drifted to his days in England when he went there to appear for the Indian Civil Service Examination. He mentioned about the other Madras men who were with him at that time. He was describing their life in Oxford and Cambridge. He said that the Ramanujan tradition was still alive in Cambridge. Then he referred to an incident which happened shortly after the *Collected papers* of Ramanujan appeared in 1927. One of the Indian students, it appears, decried some of the details found in the biography of Ramanujan. He said that Indians were being brought into disrepute by recording such cock and bull stories in a serious biography. He apparently referred to the trans-rational information mentioned in Sec B21. I said that one duty of a biographer was to make a faithful record of the facts collected. It would be unscientific on his part to reject certain facts because of his own belief or disbelief in them. This student of mine went on defending the attitude of his colleagues among the ICS Probationers in Cambridge. He finally said that men like Prof Seshu Ayyar and Dewan Bahadur Ramachandra Rao should not have published such details. I felt obliged to exonerate these gentlemen. Therefore, I disclosed those who had drafted the biography. Then my old student felt apologetic and said, "If I had known that you were its author, I would not have been so vehement about it (!) I was only telling you the fuss that one of my friends in Cambridge made about this. Some of us in the group did not see eye to eye with him. However, I now realise that such pieces of information should be recorded, leaving their evaluation to others."

B4 Evidence of Friends

Some intimate friends of Ramanujan have been mentioning quite often about Ramanujan's belief in occult phenomena. Sitting on the sands of the Triplicane Beach, he used to narrate occult experiences. He also used to give occult accounts of the stars and the other heavenly bodies. He would give occult interpretations of our epics. Those listening to him were very much impressed by the sincerity and certitude with which he gave such occult expositions.

B5 Posthumous Phenomenon

I may here refer to an even more uncanny occult happening. It was 1934. I was with K S Krishnaswami Ayyangar, a leading advocate of Madras and later a judge of the High Court. He had known Ramanujan as a boy in Kumbakonam. With the help of the Ouja Board, he invoked Ramanujan. There was response. I asked him. whether he was continuing his mathematics. He replied, "No. All interest in Mathematics dropped out after crossing over." I asked him what he was occupying himself with. The reply was, "Meditation. Studying *Vishnu-sahasra-nama*" (a hymn of 1,008 names of God). I then told him about his third Notebook which was in my possession in the University Library. Only a few pages of it had been filled up. It was all a series of tabulated numbers. Nobody could make out what it was about. Could he throw any light on it? Ramanujan replied, "I do not remember about it. Bring the Notebook to this place this time next week. I shall look into it. If I can recollect anything, I shall tell you." The Notebook was taken to Krishnaswami Ayyangar's house at the appointed time. Ramanujan was invoked. He asked me to turn the pages slowly. Then came the reply, "I now remember. I was working on Mock-Theta Function." This brought to my mind •Prof G H Hardy's reference to Ramanujan starting work on the Mock-Theta Function, before he left Cambridge in 1919.

B6 Psycheo-Genetic Force

I had completed the draft biography of Ramanujan by December 1923 for inclusion in the *Collected papers*. On 4 January 1924, I took charge of the University Library in Madras. Two months later, Seshu Ayyar asked me to meet Dr MacPhail, the Vice-Chancellor, and get the draft biography finalised. It was March 1924. The Vice-Chancellor read with me my draft of Ramanujan's

16

biography. Having once gone through the whole text, crossing the t's and dotting the i's here and there, he came back again to the first paragraph. This paragraph had a touch of speculation. It referred to the appearance of men of genius in unexpected families. It referred to the father of Ramanujan having been a petty clerk in a cloth merchant's shop, without any contact with mathematics beyond the lowest level of lower arithmetic. It also referred to the least expected time and circumstance, of the appearance of a genius. It mentioned that there was, at the turn of the twentieth century, hardly any extent of any tradition of higher mathematics in India. For, it had gone into a state of cultural exhaustion and sleep before modern mathematics took shape in Europe. The paragraph ended with this speculation: "It is not possible to explain the phenomenon of Ramanujan except on the hypothesis of the ever-increasing *Purvajanma-vasana* — the Psycheo-genetic force — gaining in momentum all through the march of a soul from embodiment to embodiment." It further stated that enough statistical data had not yet accumulated about the frequency of appearance of such phenomenal men of genius, and of the social and hereditary factors bearing on them, to make possible the formulation of any empirical law regarding the subject. Dr MacPhail read it, looked at me with a smile, and said, "I find in this paragraph the fact of your being a Hindu and a student of mathematics and particularly statistical science. As a historian, I am concerned only with facts — well-established facts. I am striking out this paragraph."

B7 Achievement in Short-life

Ramanujan was born at Erode in Madras Presidency on 22 December 1887. He died in Madras on 26 April 1920. In this short span of thirty-two years and four months, Ramanujan attained great eminence as a gifted mathematician. His reputation was not confined to Tamil Nad. Nor was it confined only to India. It spread throughout the world wherever mathematics was valued. His reputation was not a fleeting one. It is one that will last as long as interest in mathematics continues.

B71 FELLOW OF THE ROYAL SOCIETY

Ramanujan was elected a Fellow of the Royal Society of London

17

even in his thirty-first year.   Few had had this honour before him.
He was the second Indian to become an F R S.

B72   PUBLISHED PAPERS

During the short period of five years extending from 1914 to
1918, twenty-six of Ramanujan's Papers were published in the
British learned periodicals and six in the *Journal* of the Indian
Mathematical Society.   They were all of great depth.   Shortly
after his death, the University of Madras decided that his *Collected
papers* should be published.   The volume came out in 1927.   It
was brought out by the Cambridge University Press.   It consisted
of 355 pages and contained all his thirty-one published papers. It
also contained the questions and solutions contributed by
Ramanujan to the *Journal* of the Indian Mathematical Society.

B8   Anecdotes

Several anecdotes have come to my knowledge while collecting
data for the biography.   Some of them occurred in Great Britain.
Some of them may be of narrative interest. They are inserted at
different places in this book.

SUPER-ACTIVITY PERIOD 1: 1907 TO 1911

C0 The Trigger Book

The book, known to have stimulated Ramanujan into mathematical ventures, was not a profound and inspiring one. It was a dry-as-dust print of the 'coaching notes' of George Shoobridge Carr of London and Cambridge. It was entitled *Synopsis of elementary results in pure and applied mathematics.* It was in two volumes published in 1880 and 1886 respectively. Its coverage ended with 1860. As stated by Hardy, the astounding edifice of analytical knowledge and discovery of Ramanujan was based on the 6,000 theorems of Algebra, Calculus, Trigonometry, and Analytical Geometry listed by Carr. If we remember that Ramanujan not only recreated in his chosen field the results obtained by Europe in the nineteenth century, but also went quite ahead of the contemporary mathematical knowledge, Carr's *Synopsis* could have been no more than a trigger to release his explosive mathematical abilities.

C1 Anecdote One: Introduction to Carr's "Synopsis"

When Ramanujan was still a pupil in the Town High School of Kumbakonam, the following incident occurred:

RAMANUJAN: What is this big book?

ELDERLY FRIEND: I showed this to you purposely.

RAMANUJAN: What is it, uncle?

ELDERLY FRIEND: Did you not master Loney's *Trigonometry* even when you were 12? Are you not the boy who has calculated the length of the equator of the earth?

RAMANUJAN: You are pulling my legs, uncle. Leave it alone. Tell me about this book.

ELDERLY FRIEND: Carr's *Synopsis of pure and applied mathematics.*

RAMANUJAN: Will you allow me to glance through this book? ...They are all very interesting.

ELDERLY FRIEND: I am glad you like it. Keep it with you for some time.

C2 School Career

Ramanujan was absorbed with this book for several days. He verified many of the results in the book. He got excited. He discovered many new results of his own. All this he did without prejudice to his school work. Indeed, in the matriculation examination of the University of Madras held in December 1903, he gained a place in the 'first class'. In those days 'first class' was rather rare.

C3 College Career

The position that he earned in the Matriculation Examination secured for him the Subramaniam Scholarship. This enabled him to join the F A (First examination in Arts) Class in the Government College, Kumbakonam. In those days the course of studies in F A was a mixed one. It had not only Mathematics but also Physiology and History of Rome and Greece in addition to English and an Indian language. But his mind was too much taken away by Mathematics. Partly due to this and partly due to family circumstances, he could not pass the college examination at the end of the first year. He therefore lost his scholarship. This made him leave Kumbakonam. He completed his second year's course in the Pachiappa College in 1906 and appeared for the University examination (F A) in December 1907. But he failed.

C4 A False Story

After his mathematical ability came to be recognised about five years later, a false story was rumoured. According to it Ramanujan was supposed to have failed in Mathematics. This story gave an easy handle for cynics to decry our University system. It even caused disaster to a remarkably gifted young mathematician who was my student — T Vijayaraghavan. He systematically neglected his legitimate work in his honours class in mathematics, thereby equating himself with Ramanujan on the basis of the above-mentioned false story.

C5 The True Story

I had an unexpected chance to discover the falsity of the story.
20

About 1922, Statistical Methods came to be prescribed by the University of Madras as a special subject in the honours course for mathematics. Seshu Ayyar was teaching the subject. I also took interest in it. We both desired to apply statistical methods to some educational problems. Ultimately we decided to study the marking system prevalent in the University of Madras. Seshu Ayyar was then a Member of the Syndicate of the University. He therefore obtained permission for me to have access to the old marks-books of the University for about fifteen years. This was in 1922. These marks-books related to the Intermediate Examination from 1911 onwards and its predecessor F A Examination during a few earlier years. For my statistical study, I had to transfer the marks to the usual 125 x 75 mm cards — of course without the mention of the names of the candidates. But the marks-books contained the names of the candidates. I found Ramanujan's F A marks in one of those volumes. He had really scored a very high percentage of marks in mathematics. His failure was due to poor marks in the other subjects. This is the true story.

C6   A Seer in Mathematics

The period 1907-11 appears to have been the first period of super-activity in the life of S Ramanujan. Inner light began to lead him. And the joy of cultivating the region of knowledge lighted up by it began to spur him on and on. The urge for the pursuit of Mathematics became irrepressible. The depression due to failure in the F A examination could not repress it. Failure to get employed could not shake it. Poverty and penury could not obstruct it. His research marched on undeterred by any environmental factors — physical, personal, economic, or social. Magic Squares, Continued Fractions, Hyper-Geometric Series, Properties of Numbers — Prime as well as Composite, Partition of Numbers, Elliptic integrals, and several other such regions of mathematics engaged his thought. But, during or earlier than that time, hardly any thought was created in the country on some of these problems. The thought created in the West had not even been disseminated in the country. Everything had to be done and discovered by him *de novo*. Ramanujan had cultivated an unusual systematic habit. Each result that he obtained he recorded

21

in a quarto notebook. Proofs were often absent. This might have been due to two causes. Probably he *saw* some of these results directly unmediated by formal proofs. Again, even where he arrived at them by laborious work, he could not find the mental set to copy all the steps in the proofs. Ideas were pouring in at a rate which militated against copying at leisure. The result has been his extraordinary Notebook — the first of his three notebooks. The profundity of the contents of this Notebook is still staggering. These contents have stimulated many workers in their respective fields. The proofs of many are yet to be worked out. The accuracy of some is yet to be established. Surely, all this could not have been seized by the intellect alone. Intuition should have played a large part in this period of super-activity. Ramanujan was indeed a *Drashta* (= a Seer) in Mathematics.

C7 Life During Super-Activity Period 1

Few details are known about the day-to-day and the year-to-year life of Ramanujan during this period of super-activity. There are anecdotes of his having had to go without food on certain days. There are other anecdotes of a kind old lady in a neighbouring house inviting him for a meal in her home in the middle of the day, when most of the members of her family would have gone away. I had known this lady. She knew nothing of Mathematics. It is the gleam in the eyes of Ramanujan and his total absorption in something — it is these that had endeared Ramanujan to her. Evidently most of this time, he was not living with his parents. Though married in 1909, his wife was too young to have joined him. Perhaps these factors were a blessing in disguise. No domestic work distracted his attention.

CHAPTER D

ECONOMIC SUFFICIENCY

D0 Demand of Worldly Factors

But such a state of absorption does not ordinarily continue long in most people. Economic pressure is inexorable. Family responsibility does exercise its own pressure. A moment comes when the break-even point is reached between care-free pursuit of a subject under inner force and the demand of worldly factors. Such a moment arrived in the life of Ramanujan in 1910. He had then reached twenty-three. He had to seek employment.

D1 On the Look-out for Employment

He looked for openings. He had heard of the Indian Mathematical Society founded by Prof V Ramaswamy Ayyar in 1907. He learnt that though he was familiarly described as Professor, he was in reality a Deputy Collector in the Madras Civil Service. As Deputy Collector, he was *ex officio* President of the Local Body of his division. Ramanujan thought of him as a possible patron, both because of his identification with Mathematics and of his possible influence as a Deputy Collector. The headquarters of Ramaswamy Ayyar was then at Tirukkovilur. Ramanujan therefore reached that place late in 1910.

D2 Anecdote 2: Interview with Ramaswamy Ayyar

RAMANUJAN: I am interested in Mathematics.

RAMASWAMY AYYAR: Is it so? Come along. Kindly take your seat. What have you done so far?

RAMANUJAN: This Notebook contains some of the theorems and results got by me.

RAMASWAMY AYYAR: Pass it on to me... Most of these appear to be new. My goodness! Whatever page I look into, I find it to be a mine of new theorems and formulae. What a feast! Where are you working?

RAMANUJAN: I am unemployed.

RAMASWAMY AYYAR: (Still turning through the Notebook) I hope you have sufficient ancestral property.

RAMANUJAN: No sir, my family is poor. My father is a petty clerk in a cloth merchant's shop in Kumbakonam. Moreover, sir, last year my parents made me enter into married life.

RAMASWAMY AYYAR: (Still with his mind buried in the Notebook) Is it so?

RAMANUJAN: Sir, be pleased to give me a clerk's post either in your Office or in the Taluk Board's Office. I can then earn my livelihood.

RAMASWAMY AYYAR: It is too bad. If you become a clerk in any of these offices, your mathematical abilities will soon disappear. I do not want to sin that way.

RAMANUJAN: Sir, you should not say like that. Who else will help me?

RAMASWAMY AYYAR: Do not think that I want to disoblige you. You will get some real help. Just wait for a few minutes.

(Then, Ramaswamy Ayyar went into his office room and wrote out a letter of recommendation to P V Seshu Ayyar).

RAMASWAMY AYYAR: Take this letter, Ramanujan, go and meet Prof Seshu Ayyar of the Presidency College, and give him this letter. Do you know him at all?

RAMANUJAN: Yes, sir. I was his student in the Government College, Kumbakonam.

RAMASWAMY AYYAR: Then it will be easy for you to meet him.

D3 Anecdote 3: Interview with R Ramachandra Rao

Prof Seshu Ayyar gave a note of introduction to Dewan Bahadur R Ramachandra Rao, Collector of Nellore District, and President of the Indian Mathematical Society. With that note Ramanujan went to Nellore in December 1910. In the words of Ramachandra Rao himself, "In the plenitude of my mathematical wisdom I condescended to permit Ramanujan to walk into my presence. A short uncouth figure, stout, unshaved, not over-clean, with one conspicuous feature — shining eyes — walked in with a frayed Notebook under his arm. He was miserably poor. He had run away from Kumbakonam to get leisure in Madras to pursue his studies. He never craved for any distinction. He wanted leisure; in other words, that simple food should be provided for him

24

without exertion on his part, and that he should be allowed to dream on."

RAMACHANDRA RAO: Prof Seshu Ayyar says you have a Notebook with you.

RAMANUJAN: Yes, sir. Here it is. May I just read out some of my theorems from it?

RAMACHANDRA RAO: No. Pass the Notebook on to me (After perusing the Notebook for some minutes). I am not so much of a mathematician to understand all this stuff.

RAMANUJAN: Here are some simple results, sir.

RAMACHANDRA RAO: Yes, they may be simple. But we do not find any of them in the books that we know. Tell me what you want.

RAMANUJAN: I want some means to earn my livelihood.

RAMACHANDRA RAO: Do not worry yourself about it. Where do you live?

RAMANUJAN: Summer House, Sami Pillai Street, Triplicane, Madras (This is the pretentious name of a small, poorly ventilated house).

RAMACHANDRA RAO: Alright. Go back to your place. You will get the necessary money to meet your needs.

D4 Anecdote 4: Talk with "Sandow"

Every month Ramanujan got a money order for his expenses. He was working all day and most of the night. His recreation was a walk on the Triplicane Beach in the cooler hours of the evening and conversation with his intimate friends. These were partly due to his shyness and partly due to his preoccupation with his own thought in Mathematics. But his unusual mathematical capacity slowly pervaded among a larger section of the people. It was February 1912. K S Srinivasan, popularly called "Sandow" by his friends and a classmate of mine in the Madras Christian College, had known Ramanujan intimately while at Kumbakonam. He called on Ramanujan at Summer House one evening.

SANDOW: Ramanju, they all call you a genius.

RAMANUJAN: What! Me a genius! Look at my elbow, it will tell you the story.

SANDOW: What is all this Ramanju? Why is it so rough and black?

RAMANUJAN: My elbow has become rough and black in making a genius of me! Night and day I do my calculation on slate. It is too slow to look for a rag to wipe it out with. I wipe out the slate almost every few minutes with my elbow.

SANDOW: So, you are a mountain of industry. Why use a slate when you have to do so much calculation? Why not use paper?

RAMANUJAN: When food itself is a problem, how can I find money for paper? I may require four reams of paper every month.

SANDOW: Tell me honestly what do you do for your food. Do you work anywhere?

RAMANUJAN: Our Prof Seshu Ayyar introduced me to Dewan Bahadur R Ramachandra Rao, the Collector of Nellore. That great man has been providing me with money every month.

SANDOW: Then why do you worry yourself?

RAMANUJAN: How long am I to depend on others? The humiliation of it has gone deep into me. Therefore I did not take the money from last month.

SANDOW: What a rash thing to do! What are you going to do now?

RAMANUJAN: I joined the Madras Port Trust Office as a clerk on the 9th of this month. Pay Rs 25 a month.

D5 Madras Port Trust

Seshu Ayyar and Ramaswamy Ayyar had been for some time appreciating Ramanujan's reluctance to live on a monthly donation from somebody. They wanted to help Ramanujan to realise his wish that he should earn his living in return for work done. They finally persuaded S Narayana Ayyar, Manager of the Port Trust in Madras, an active worker in Mathematics, the Treasurer of the Indian Mathematical Society, to take Ramanujan in o his staff. At their suggestion, on 9 February 1912, Ramanujan applied to the Chief Accountant of the Madras Port Trust for appointment as a clerk. The Office Note recommending his application said, "He is reported by M G R to be a Mathematical Genius. Mr Middlemast says of him to be 'exceptionally intelligent in Mathematics'. Sanctioned". Accordingly, the following Appointment Order (A 5345) was issued on 9 February 1912 — "With reference to his application dated 9 February 1912, S

Ramanujan is informed that he has been appointed by the Chairman as a clerk in this office on Rs 25 per mensem. He should report himself to duty on the first of March 1912." As a result, Ramanujan had the satisfaction of getting a clerk's post for earning his own living — a wish that he had been pursuing for some years.

D6 Anecdote 5: Madras Port Trust

Dame rumour circulated in those days a plausible piece of information. Whether it was genuine or apocryphal, it is worth stating. Some sheets of paper containing some results in Elliptic Integrals were found mixed up in a file reaching Sir Francis Spring, the Chairman. Immediately he sent for his Manager, S Narayana Ayyar. He pretended to be angry at the Manager for using office hours for private work. The Manager felt puzzled and pleaded not guilty. Sir Francis showed him the sheets with Elliptic Integrals and said, "you are the only person in my office who dabbles in this kind of Mathematics." The Manager said that it was not his handwriting. "Who else dabbles in higher mathematics?" asked Sir Francis. The Manager mentioned Ramanujan's name. Then Sir Francis had a hearty laugh saying, "I know it. I merely teased you."

D7 Continued Interest of Madras Port Trust

Ramanujan worked in the Madras Port Trust for only about one year. But Francis Spring and S Narayana Ayyar continued their interest in him all through his life. As shown in Chapter E, they actively worked to get him the quiet and the facilities normally going with life in a University. Later, they followed his career in the Universities of Madras and Cambridge, with persistent interest. A considerable quantity of Ramanujan Papers had thus accumulated in the Madras Port Trust.

D8 Ramanujan Archives

In 1950, Mr M S Venkataraman, son-in-law of S Narayana Ayyar and formerly Traffic Manager of the Madras Port Trust, was the Chairman of the Port Trust. On 19 June 1950, he requested Mr K Santhanam, Minister of State for Transport and Railways, Government of India, to have the Ramanujan Papers preserved in the Indian National Archives. On 14 July

1950, the Government of India agreed to take over the Ramanujan Papers. The Papers were transferred to the Indian National Archives on 22 July 1950. According to a Press Release of 17 October 1950, "The file consists of 103 folios of a variety of sizes. It commences with his application of February 1912 to the Madras Port Trust for the post of a clerk on Rs 25 a month and closes with the report of his untimely death at the age of 33. The file contains some of his abstruse mathematical observations which exhibit his amazing flair for arriving at the correct solution of complex abstract problems. Steps for its preservation in the National Archives are being taken, while a micro-film positive copy will be supplied to the Madras Port Trust for record." Other sections of Ramanujan Archives are in the University of Madras, Cambridge University, and the Royal Society (London).

UNIVERSITY SETTING

E1 Efforts to Secure Research Scholarship

The Indian mathematicians mentioned in Chapter D wished to secure a Research Scholarship for Ramanujan. Accordingly R Ramachandra Rao had asked C L T Griffith of the Madras Engineering College to do what he could about Ramanujan. On 12 November 1912, Griffith wrote to Prof M G M Hill of the University College (London), for his opinion on some of the results got by Ramanujan. He replied that the "intuitive results" might be of interest.

E2 Action by Gilbert Walker

On receipt of this opinion of Hill, Francis Spring took the action described in his own words as follows:

"On 25 February 1913 Dr Gilbert Walker, F R S, himself a firm mathematician and Senior Wrangler, now Head of the Indian Meteorological Department, paid a visit to the harbour in connection with our tidal observatory. I took the opportunity of bringing the work of S Ramanujan to his notice through Mr S Narayana Ayyar, himself a mathematician and Hon Secretary to the Indian Mathematical Society.

"Dr Gilbert Walker disclaimed himself ability to judge some of Mr Ramanujan's work and said that Mr Hardy of Trinity College, Cambridge was in his opinion the most competent to arrive at a judgement of the true value of the work.

"Mr Ramanujan had already been in correspondence with Mr Hardy, a letter of whose dated 8 February 1913—just 17 days before Dr Walker's visit here—is in this file."

On 26 February 1913 Gilbert Walker wrote to the Vice-Chancellor of the Madras University, who was a British Judge of the Madras High Court, asking him to give a Research Scholarship to Ramanujan.

E3 Academic Support of the Board of Studies

On 13 March 1913, B Hanumantha Rau, Professor of Mathematics in the engineering college, wrote to S Narayana Ayyar as follows:

"I am calling a meeting of the Board of Studies in Mathematics in the Senate House at 5-30 p m on Wednesday the 19th instant for the purpose of considering what we can do for S Ramanujan. Will you kindly come over and assist us? You have seen some of his results and can help us to understand them better than the author himself." The Board of Studies resolved to support the recommendation of Gilbert Walker.

E4 Legal Hurdle

With the support of the Board of Studies in Mathematics, the Vice-Chancellor placed the matter before the Syndicate, for consideration. He personally backed it as an extraordinary case which should be met in an extraordinary way. He suggested that Ramanujan should be awarded a research scholarship. However, some Members of the Syndicate put a spoke in the wheel. They cited the university regulations on research scholarship, which prescribed the possession of a Master's Degree as a necessary qualification. For some time, it looked as if the Vice-Chancellor's proposal would be turned down. While in this predicament, Justice P R Sundaram Ayyar showed a way out. He pointed out that, as stated in the Preamble to the University Act, one of the functions of the University was to promote research. Here was a candidate of proven phenomenal capacity for research. The University would be failing in its duty if it did not use him as a means for the promotion of research. The regulations were University-made. Any regulation obstructing the implementation of the provisions of the Act of the Legislature should be deemed to be *ultra vires*. This cleared up the position. The legal hurdle was crossed. In abundance of caution, the University got the consent of the Chancellor of the University to give a special research scholarship to Ramanujan.

E41 Historic Minute

The following is the text of the historic minute of the Syndicate of the Madras University on the subject.

"Read letter, No. 1-12, dated 26 February 1913, from the Director-General of Observatories, Madras, bringing to the notice of the University the original character of the work in Mathematics produced by S Ramanujan, Clerk in the Port Trust Office, and stating that the University will be justified in enabling him for a few years to spend his whole time on Mathematics without any anxiety as to his livelihood.

"Read also letter, dated 25 March 1913, from the Chairman, Board of Studies in Mathematics, stating that the Board recommends the award of a scholarship of Rs 75 per mensem to S Ramanujan for a period of two years on certain conditions.

"Resolved that the recommendation of the Board be accepted and that the sanction of the Government to payment of the proposed scholarship from the University Fee Fund be requested in accordance with the provisions of Section XV of the Act of Incorporation and Section 3 of the Indian Universities Act, 1904."

E42 QUARTERLY REPORT
The condition imposed was that Ramanujan should submit a quarterly report on his work (*See* Sec B2 for an account of the first quarterly report). Ramanujan sent three quarterly reports before leaving for Cambridge early in 1914.

E43 UNIVERSITY OF MADRAS: THE FIRST ACADEMIC SETTING
Ramanujan was thus enabled to give up his clerical drudgery in the Madras Port Trust. He joined the University of Madras as a Research Scholar on 1 May 1913 on a stipend of Rs 75 per month. This gave him just the facility needed by a man of his ability and temperament. He was put above want. He could dream on and make his flight in the ethereal regions of higher mathematics free from worry and care of any sort. The Madras University Library had just then begun to function. Trivikrama Rao was then in charge of it. It was housed in a wing of the Connemara Public Library Buildings. It had a few alcoves to its share. Ramanujan used to spend much of his time in one of these alcoves. This was the first academic setting secured for Ramanujan. During this period, Ramanujan used to spend the morning hours and the early hours of the night in the house of S Narayana Ayyar. They would be working together on Elliptic Integrals or other problems in Mathematics, till late in the night. Each would write his ideas on a slate to help discussion.

E5 Efforts to Secure Overseas Scholarship

After the research scholarship was secured in the University of Madras, the efforts of all who had taken part in it, were turned on sending Ramanujan to Cambridge. For, it was felt that the academic setting of the right standard to help Ramanujan to reach his own fullness, was not available in India at that time.

E6 Initiative of G H Hardy

E61 FIRST LETTER TO HARDY

At the suggestion of his well-wishers, in January 1913, Ramanujan wrote his first letter to G H Hardy, Cayley Lecturer in Mathematics in Cambridge. This letter has been published in several books. The turn of expression in that letter justifies one in inferring that the letter should have been largely worded for him by the seniors who had been helping him. Secondly, "the-self-boasting" obliquely implied in it is quite un-Ramanujan-like. For, the shyness and humility of Ramanujan were un-exampled. The correct term to describe this quality of Ramanujan is the Sanskrit word *Hri* in the sense in which Valmiki used it in evaluating Rama. The enclosure to the letter contained 120 theorems and formulae. These made Hardy marvel at the extraordinary mathematical powers of Ramanujan.

E62 CALL OF CAMBRIDGE

The result was that Hardy immediately decided that Ramanujan should no longer be left in mathematical solitude in India, but should be brought over to work with his peers in Cambridge. I can very well visualise the spontaneity and certitude with which he should have acted on this decision of his. For, in March 1925 I personally saw him make a similar decision with spontaneity and certitude. He was then Savilian Professor of Geometry in the New College, Oxford. I met him at that time. I attended one of his lectures. I also had lunch with him. My main purpose was to place before him a paper on the Tauberian Theorem by T Vijayaraghavan who was a gifted student of the Department of Mathematics in the Presidency College, when I was teaching there. After perusing this paper, Hardy stood up with a jerk and said, "This is a remarkable mathematician from your country — next only to Ramanujan. He should be brought over to me." This he said, even without waiting to know any details about Vijaya-raghavan. I told him that he had failed in the Honours Examination just the previous year. Hardy asked how it happened. I said that Edward B Ross—my own professor and one of his own old students — and myself were examiners and that in spite of our personal knowledge of his extraordinary mathematical ability we could not give him even pass marks, as he had not answered most of the questions in all but one paper, *viz.* the one on Functions of a Complex Variable. After hearing this, Hardy remarked, "It makes no difference whatever". He immediately wrote out two copies of memorandum, gave them to me, and asked me to send them to whomsoever I liked. The memorandum emphasized that Vijayaraghavan should be sent over to Oxford without any delay. I sent one copy to Dr MacPhail, the Vice-Chancellor of the University of Madras and the other to Sir A P Patro, the Education Minister of the Government of Madras. Eventually Vijayaraghavan became the first professor of the Ramanujan Institute established by Alagappa Chettiar in Madras in 1949 (*See* Sec N31).

E63 HARDY'S FIRST ATTEMPT

In the case of Ramanujan, Hardy wrote to the Secretary for Indian Students, in the India Office, London, suggesting that some means should be found to get him to Cambridge. This suggestion was passed on by that officer to the Secretary of the Students' Advisory Committee in Madras.

E64 HARDY'S SECOND ATTEMPT

Hardy's keenness to have Ramanujan in Cambridge continued unabated. Mr E H Neville, a young Cambridge mathematician, was to come to the University of Madras as Visiting Lecturer on Differential Geometry in the winter months of 1913-14. Hardy took advantage of this. He commissioned Neville to do everything possible to secure the residence of Ramanujan in Cambridge.

E7 University of Cambridge: The Second Academic Setting

Neville's presence in Madras created a favourable atmosphere. I remember that "Ramanujan" was the topic of the day in Madras at that time (*See also* Sec B1). The letter of Hardy to the Secretary for Indian Students in the India Office (London), was brought up for consideration by the Syndicate of the University of Madras. The University rose to the occasion. It sanctioned without any opposition from anybody, an annual overseas scholarship of £250 for two years and also the necessary amount for travel and outfit expenses. The period of scholarship was later extended to five years.

E8 Initiative of Francis Spring

At this stage Sir Francis Spring spoke to the Governor of Madras and followed it up with a letter to the Private Secretary, from which the following are extracts:

E81 TRANSCENDENTAL ORDER OF GENIUS

"If I understand right, His Excellency has the Educational portfolio. So I am anxious to interest him in a matter which I presume will come before him within the next few days — a matter which under the circumstances is, I believe, very urgent. It relates to the affairs of a clerk of my office named S Ramanujan who, as I think His Excellency has already heard from me, is pronounced by very high Mathematical Authorities to be a Mathematician of a new and high, if not transcendental, order of genius.

E82 MANDATE FROM CAMBRIDGE

"During the last 8 or 9 months various Mathematicians in the first rank in Cambridge, Simla, and Madras have had before them selections from his work and have pronounced upon them in terms of the very highest eulogy. Just now, as probably His Excellency is aware, Mr Neville, who I think, is a Senior Wrangler and a Fellow of Trinity, Cambridge, has been in Madras giving a series of lectures on certain phases of the Higher Mathematics to Honours students and others interested. Under a mandate from Cambridge he has interested himself greatly in Ramanujan and there is every reason to hope that he may be persuaded to go to Cambridge for a year or two so that under expert guidance, not only may the fruits of his genius be given to the world but also we may hope, his own fame, future usefulness and personal prosperity may be secured — matters probably quite impossible if he remained in a backwater like Madras for the rest of his life."

E83 CONFIRMATION OF OVERSEAS SCHOLARSHIP

"I now come to the point where His Excellency may perhaps be able to interfere with advantage. Last evening I learnt from Mr Littlehailes and others that the University Syndicate had decided, subject to sanction of Government, to set aside a sum of Rs 10,000 in order to secure Ramanujan's visit to England for a couple of years. Messrs Littlehailes and Neville begged me to intercede with His Excellency with a view to the speedy confirmation of this action of the University Syndicate. But I wish to make it quite clear that I write under no mandate from the Syndicate but merely as a private individual interested in my own employee, Ramanujan, as well as in Mathematics. Mr Arthur Davies will doubtless arrange for the voyage to England and that Ramanujan's orthodoxy may be maintained unimpaired. . . .

"I myself am very far from being mathematician enough to express adequately what has been said to me by several who are fully qualified to express an opinion on the subject of the potential value to Science of the new line of thought in which Ramanujan's investigations lie. I am assured however by those who ought to know what I am talking about that they may conceivably be epoch-making and as such well worthy of financial support at the hands of the Madras University."

This brought the following reply dated 5 February 1914 from the Private Secretary to the Governor.

"Your letter of 5 February. His Excellency cordially sympathises with your desire that the University should provide Ramanujan with the means of continuing his researches at Cambridge, and will be glad to do what he can to assist in the proposal."

RAMANUJAN'S LIFE ABROAD

F1 Ramanujan Groomed for Cambridge Life

F11 CONSENT TO GO TO CAMBRIDGE

The first step in the preparation for Ramanujan's voyage was to shake off the obstruction of orthodoxy and social custom, then prevailing, against crossing the seas. Seshu Ayyar appealed to Ramanujan's mother. Ramaswamy Ayyar appealed to Ramanujan. So did Ramachandra Rao. What was of even greater weight to Ramanujan was the persuasion by M T Narayana Ayyangar, Professor of Mathematics in the Central College, Bangalore, Editor of the *Journal* of the Indian Mathematical Society, and a Member of the Board of Studies. This gentleman had very orthodox ways. In spite of it, he told Ramanujan that orthodoxy about crossing the seas should give way in the interest of Mathematics. At long last, Ramanujan's mother yielded and Ramanujan agreed to go to Cambridge (*See* Sec B23).

F12 PREPARATION FOR LIFE IN ENGLAND

Then came the problem of outfit for life in England. During that period, Littlehailes used to ride his motor-cycle with legitimate pride, with Ramanujan in the side car to help him in getting his outfit made. Both he and Neville initiated Ramanujan in slow degrees into the way of life prevailing in England. Ramanujan had to learn to wear European clothes. The then South Indian mode of growing a long tuft of hair became a problem. Ramachandra Rao persuaded or forced — one cannot say which — Ramanujan to conform to the English mode. But nobody could succeed in making Ramanujan give up his vegetarianism. These details may now appear to be too trivial. But they were serious problems in those days for any normal young man of the Hindu-fold.

F13 FAMILY ALLOWANCE

Then came his duty to his parents. To discharge this duty

Ramanujan asked the University of Madras to pay his parents
an annual sum of £50 out of his scholarship.

F14 FOOD ON VOYAGE
On 11 March 1914 Francis Spring wrote to the Steamer Agents
asking them to provide Ramanujan with vegetarian food throughout
the voyage.

F15 VOYAGE BEGINS
Ramanujan was thus well groomed for his life in England by
his friends and well-wishers. He carried with him all the good
wishes of the University of Madras. He sailed from Madras on
17 March 1914 on board the S S Nevasa.

F16 RAMANUJAN ARRIVES IN CAMBRIDGE
The Madras Dailies of 13 May 1914 had the following
news:

"Mr S Ramanujan of Madras, whose work in Higher Mathematics has excited the wonder of Cambridge, is now in residence
at Trinity. He will read mainly with the two Fellows of the College
— Mr Hardy and Mr Littlewood. They are going through
masses of work he has already done, and are making some sur-
prising discoveries in it!"

F2 Ill at Ease
Ramanujan carried with him the best wishes of the mathematicians
and other friends in Madras. He was received with acclamation
and admiration in Cambridge. In spite of all this, Ramanujan
seems to have been ill at ease both during the voyage and his stay
in England. I came to know about it early in 1925. One day in
January that year, I went to the University College at Reading
to study the working of its library. Members of the teaching
faculty of that college invited me for lunch. I found Neville
among them. On seeing me and my turban, his old memory of
Madras seemed to have come back to him. He greeted me and
took his seat at the table by my side. He found out from me that
I was a post-graduate honours student of Mathematics in Madras,
when he came there as a Visiting Lecturer. This recollection of

the time, place, and subject urged him to invite me to his home that evening. We had a long reminiscential talk.

F3 Turban Problem

Mrs Neville told me that wearing a hat was a torture to Ramanujan during the voyage as well as during the whole of his stay in England. She then asked me, "Why did your people force Ramanujan to wear a hat instead of a turban?" I replied that it should have been due to a mesmerised mentality of the people of a Dependency or to a blind belief that a person walking in the streets of London or Cambridge with a turban on would be hooted down or even stoned by the mob. She asked me if I had experienced it. I then said, "I told you that it was only a blind belief. I never had any awkward experience here. I remember only two occasions when my turban produced any visible effect at all on the people here. One was at the Hyde Park Corner on a Sunday evening. A political speaker was standing on an improvised platform and haranguing the audience on Independence for Ireland. While referring to the exploitation of Ireland by England, he pointed to me and said, "Here is a friend from India. He can understand how England can persecute a slave nation." The second occasion was when I was travelling by train from London to Croydon. The train was moving at a slow speed on account of repairs to the permanent way. I was occupying a seat near a window. The gang-coolies were referring to me as Mr A. This happened because at that time a turbaned Indian prince was involved in a law suit. The newspapers used to call him Mr A for protocol reasons."

F4 Tuft Problem

Then Mrs Neville told me that what caused Ramanujan even greater pain was his having to remove his tuft of hair before sailing for England, and asked me if I had to do so. I said "No". She asked me, "Are you still having your tuft?" I thought she would not believe me if I merely said 'Yes'. And so I took off my turban and put it on the table for a while. Mrs Neville was very much moved. She said, "Why in the world should they have forced Ramanujan to remove his tuft? He could have come as you have done. He would have been ever so happy."

38

F5 Food Problem

Mrs Neville told me that unfortunately World War I broke out within a few months after Ramanujan's arrival and that food became a serious problem—particularly, fruits and vegetables. She said that Ramanujan used to cook his own food and that as he was often absorbed in Mathematics he would cook only once in a day or two. I told her that the conditions had changed since then and that there was an appreciable number of vegetarian hotels in London and that some houses in Cambridge, London, and Oxford were also able to provide vegetarian food—even Indian dishes. "It is a great pity that such facilities did not develop in England when Ramanujan was here," she said with some sadness (*See* Secs ML 2, MN 4 and MT 3).

F6 Dress Problem

Mrs Neville also told me that the wearing of European dress was another problem for Ramanujan. He was always conscious of his outlandish dress. He could not reconcile himself to wearing stockings and shoes. Mrs Neville said, "Since you have your Indian hair-dress, I wonder if you have also brought your other Indian clothes with you." I said that I had brought my *dhoti* and that I generally wore it at home. She asked me if the land-lady or any of the inmates of the home objected to it. I replied in the negative. She said, "This is another matter, trivial though it be, in which Ramanujan could have been saved so much pain."

F7 Unfamiliar Climate

Mrs Neville told me that Ramanujan found the unfamiliar climate of England rather trying. Evidently he could not stand the cold. In spite of his mind being often tied up with his mathematical research, he did undergo a good deal of physical discomfort particularly during the winter months (*See* Sec MN 2).

SUPER-ACTIVITY PERIOD 2: 1914 TO 1918

G1   Reciprocal Learning

Naturally it took a few weeks for Ramanujan to accustom himself to the new surroundings. The kindness of Hardy and Littlewood shortened this period considerably. The one hundred and twenty theorems sent to Hardy by Ramanujan on 16 January 1913 had already been sifted carefully before Ramanujan's arrival. This sifting had disclosed the strong points in Ramanujan's mathematics and the weak ones too. As Hardy puts it, "The limitations of his knowledge were as startling as its profundity. Here was a man who could work out Modular Equations and theorems of Complex Multiplications to orders unheard of, whose mastery of Continued Fractions was on the formal side at any rate beyond that of any mathematician in the world, who had found himself the Functional Equations, the Zeta Function, and the Dominant Terms of many of the most famous problems in the Analytic Theory of Numbers; and he had never heard of Doubly Periodic Function or of Cauchy's theorems, and had indeed the vaguest idea of what the Function of a Complex Variable was." Hardy decided that it would be wrong to ask Ramanujan to submit to systematic instruction—learning Mathematics from the beginning once more. He felt that such a course would break the spell of his inspiration. At the same time it was impossible to allow Ramanujan to remain in ignorance of his own mistakes. He therefore tried to work with him in a participative way, involving reciprocal learning.

G11   HARDY'S UNCONVENTIONAL WAY

Indeed, Hardy wrote, "So I had to try to teach him, and in a measure I succeeded, though obviously I learnt from him much more than he learnt from me... He was never a mathematician of the modern school and it was hardly desirable that he should become one... And his flow of original ideas showed no symptom of abatement." This unconventional way of Hardy, I had occasion to witness during my visit to him in March 1925. As stated in

Sec E62, Hardy decided in a trice to get T Vijayaraghavan to Cambridge. He asked me to meet the Secretary of the Oriental Delegacy in Oxford and get Vijayaraghavan admitted to the University. The Secretary was a retired officer of the Indian Civil Service. He proved obstructive. He ridiculed the idea of one who had failed to get a degree in India to be admitted for a higher course in Oxford. I pleaded with him that he was an exceptional student and that I knew his abilities quite well as he was my student, in Madras. But the Secretary was not only unmoved but even became insolent. I reported this matter to Hardy. Hardy immediately wrote a note and asked me to take it to the Secretary. That note referred to an Indian mathematics student who had been admitted by the Delegacy to work with Hardy. If there was any difficulty in allowing more than one student to work under him, Hardy wrote that he would have Vijayaraghavan though he had failed in the Indian University examination, in preference to the other one who had a first class. This note did the work. The Secretary admitted Vijayaraghavan murmuring against both Hardy and myself.

G2   Getting Established

Till the war broke out in August 1914, Hardy, Littlewood, and Ramanujan worked incessantly on several of the results conjectured by Ramanujan. Ramanujan picked up the modern methods of vigorous proof in Mathematics. He also began to establish by vigorous proof several of the results conjectured by him while in India. Some turned to be right and a few were seen to be wrong. This did not disturb him, for the joy was of the same order whether a conjecture proved to be right or wrong. Littlewood had to leave this group on war duty. But, Hardy and Ramanujan marched on in a hectic way from the summer of 1914 to the summer of 1917.

G3   Output while in England

Only five of the thirty-seven collected papers included in his *Collected papers* had been published in the *Journal* of the Indian Mathematical Society before he left India. Most of the remaining thirty-two papers took shape during the super-activity period 1914-17, though many of them were published only in slightly later years. Here is a chronological distribution of the papers of Ramanujan published while in England.

Year	Number of papers published	Cumulative total
1914	1	1
1915	9	10
1916	3	13
1917	7	20
1918	4	24
1919	4	28
1920	3	31
1921	1	32

It can be seen that about 60 per cent of the papers were actually released during the hectic period. Of the thirty-two papers, seven were written in collaboration with Hardy. Apart from this, Hardy had edited practically all the papers.

G4 Proposal on Collected papers of Ramanujan

Shortly after the death of Ramanujan, P V Seshu Ayyar desired that the *Collected papers* of Ramanujan should be published. This, he discussed with the Members of the Board of Studies in Mathematics of the University of Madras. It was felt that the inclusion, in the *Collected papers*, of the unpublished results found in the Notebooks of Ramanujan would delay the publication inordinately. Therefore, it was decided that the *Collected papers* should contain only the published papers. Accordingly, the University appointed G H Hardy, P V Seshu Ayyar, and R Ramachandra Rao to take up the task as the Editorial Committee.

G5 Fulfilment of the Proposal

When I met Hardy in Oxford in 1925 (*see* Sec L2), the question of the *Collected papers* of.Ramanujan turned up at one stage.

HARDY: Is the press copy of the *Collected papers* ready?
RANGANATHAN: No, not at all. It has not even been begun. People in Madras are under the impression that you are preparing it.

HARDY: Is it so? I did not know that. I am rather pressed for time. However, it is a piece of work that I ought to do. Do you think that your University will allow me to take as my collaborator Mr B M Wilson?

RANGANATHAN: I am sure they will allow it. You have only to write to them about it.

HARDY: If they allow it, my work will become easy.

I alerted P V Seshu Ayyar about this. The University accepted Hardy's proposal. The editing of the *Collected papers* went on vigorously thereafter. Within less than two years — in 1927 — the *Collected papers* was brought out by the Cambridge University Press.

PRECIPITATION OF HONOURS

H1   Fellow of the Royal Society

On 28 February 1918, Ramanujan's name was proposed for the Fellowship of the Royal Society at the early age of thirty. It was accepted. It has been said that this was the first occasion in modern times when a person got elected to the Royal Society at the first proposal. Niel Bohr is said to be the only other so elected.

H2   Fellow of the Trinity College

On 13 October 1918, Ramanujan was elected a Fellow of the Trinity College, Cambridge. He was the first Indian to be so elected. This fellowship carried an honorarium of £250 a year for six years. No duties or conditions were attached to this honorarium.

H3   Honour by the Madras University

Hardy suggested that the University of Madras should make a permanent provision for Ramanujan, which would leave him free for research. The University rose to the occasion. It granted him an allowance of £250 a year, for five years from 1 April 1919, the date of the expiry of the overseas scholarship he was then drawing. This award was also unconditional. The University further agreed to pay all his expenses for travel between India and Europe during this period.

H4   Felicitations by the Indian Mathematical Society

This news reached the Second Indian Mathematical Conference held in Bombay in January 1919. I remember the event as I attended that Conference. In fact, the announcement was made in a morning session in which I read my paper. R P Paranjpye—a Senior Wrangler of Cambridge University and Principal of the Fergusson College, Poona — and later knighted — was then in the chair. A message of felicitations was voted to be cabled to Ramanujan who was still in England. V Ramaswamy Ayyar went into raptures. He was the first mathematician to assess the unusual mathematical powers of Ramanujan. He was also the Founder of the Indian Mathematical Society (*See* Secs D1 and D2).

H5 Professorship of Mathematics

Littlehailes attended that Bombay Conference. He was all enthusiasm. On his return to Madras, he became the Director of Public Instruction, Madras. One of his first acts in that capacity was to move the Madras University to create a University Professorship of Mathematics and offer it to Ramanujan (*See* Sec L1). But, alas! before this intention could become a fact, Ramanujan passed away.

RAMANUJAN'S HEALTH FAILS

J1 Emotional Strain

As stated in Chap F, Ramanujan's emotional sheath was subjected to a considerable strain during the years of his stay in England. It was the result of two factors. The first factor was due to the period when he was living. At that time, social opinion in Indian communities—particularly in South Indian Communities—had not emancipated itself from the effect of centuries of their cultural exhaustion and sleep: There was much of insularity. The scale of values was of a low order. Undue value was attached to forms and rituals. Emotion got easily ruffled if there was any non-conformity to any form of ritual. There was very little travel even within the country to enable people to see equally good life being lived and equally good results being obtained under other forms and rituals. Few boys at that time would have travelled beyond a radius of 50 miles from their homes. This radius marked out usually an area where there was uniformity of outlook and scale of values in respect of matters affecting emotion. It is true that Ramanujan shot out as far as 500 miles from his home when he ran away to Vizagapatnam in an impulse. But he did not stay long enough to see sufficient difference in the scale of values in his own district and in that far-off place. Even if he had opportunity to see the difference, it would not have been sufficiently pronounced to have helped him to re-evaluate his scale of values already embedded in his emotional sheath.

J11 Strangeness of English Scale of Values

He was suddenly thrown into English Society whose emotional values were of a different kind. For example, he did not find the same visible emotional attachment between parents and their adult children. They were not living together. There was no joint-family system. The religio-emotional attachment characterised by annual ceremonies offered to the deceased parents in Indian societies was not found in England. The annual ceremony day — almost a whole week — was a period of austerity and devoted thought turned on the deceased parent culminating in a ritual spread over some hours. The disembodied soul of the father or the mother was invoked and offerings were made with fervour and sincerity. There was nothing of this kind in the English society. Now that we are half a century from that period, some of the young Indians may not be able even to visualise the profound value attached to these annual ceremonies in the days of Ramanujan. This strangeness of cultural environment caused emotional strain to Ramanujan. The suddenness in the change of cultural environment caused misery to Ramanujan. Further, there was hardly anybody in Cambridge with whom he could intimately talk about his feelings in such matters and thus get some relief (*See* Sec B3).

J2 Physical Strain

Again, food habit became a source of physical strain. With a person who has always been living with his own family or his own intimate relatives and friends in a homely atmosphere, his whole bodily system gets more or less rigidly fixed to certain food habits. His taste for food, his hours for food, and all such matters get fixed and become rigid. So long as he does not go out to a society where all these are different, he does not feel what a terrible hold the food habit has on him. It is not merely the case of emotional attachment to certain kinds of food, it is the glandular system and the physiology in general that get upset. In the years Ramanujan was in England, he could not taste his habitual Indian food even occasionally. War conditions made things worse. But by the time I was in England — in 1924-25 — there was a slight improvement. There was at least one shop in Piccadilly, from where one could get some Indian food materials. A family or two in London had learnt to cook food in the Indian style. Six of us did train a landlady in this matter. Our home was in 142 Elgin Avenue, Maidavale. It was the resort of several Indian students for about ten years thereafter — till Mrs Peterson died — tasting her Indian food once in a few months that gave them considerable relief. Several Indian students had written to me during those ten years about the food-solace they received from that home. Today things are even better. There are many Indian restaurants. There are even specialised Indian restaurants giving Malabar food,

Tamil food, Gujarati food, Bengali food, and so on. What is more important, there are several Indian shops selling Indian food materials. Therefore, a young man going to England today does not have to undergo as much strain in respect of food as Ramanujan had to in his days. It has been said that due to his non-adjustability in respect of food habit, his having to cook his own food, and his not having sufficient variety of Indian food materials such as he was accustomed to in India, he was continuously under-nourished (*See* Sec F5).

J3 Fullness of Intellectual Life

With the net sum of £200 received from the University of Madras and an Exhibition of £60 received from Trinity, Ramanujan found his annual income ample; indeed he could not spend the whole of it. For the first time, he found himself in a comfortable position. Therefore, in spite of the emotional and physical strain described above, Ramanujan lived his intellectual life with great zest in the company of the eminent mathematicians in Cambridge. Perhaps, the super-activity induced by congenial mathematical surroundings had been drawing more energy out of his body than it could afford to give. Perhaps, the emotional disturbance was not quenched but was driven underground. It should have been eating one part of him unawares. And his under-nourishment should have corroded into his physical body. This predominance of intellectual life over physical and emotional life had evidently produced an illusion in his own mind and in the minds of his co-workers that all was well with his health. For, there was no report of his body having given way to the hidden strain — emotional and physical — till the middle of 1917.

J4 Suspected Illness

In May 1917, the University of Madras heard from Hardy that it was suspected that Ramanujan had contracted an incurable disease. It was felt that he should immediately go to the tropics. However, there were difficulties in the way. Firstly, the submarine havoc made sea voyage risky. Secondly, India had been depleted of able English medical practitioners on account of the demand of war. Therefore, it was decided that he should stay in England in spite of unhelpful factors in respect of physical and

emotional comfort. He went into a nursing home at Cambridge
in the summer of 1917. He moved to sanatoria at Wells, at
Matlock, and in London. By the end of 1918, it was definitely
known that tuberculosis had set in.

J5 Return Home

Knowing that the climate of England was retarding his recovery,
he left England on 27 February 1919 on board SS Nagoya. He
arrived at Bombay on 27 March 1919. Ramanujan's mother and
one of his brothers received him in Bombay. He was disappointed
that his wife had not been brought there. After three days rest
in Bombay, he reached Madras on 2 April. His wife joined him
there. His health was then quite bad. His body had become
very thin and emaciated. It made his family, friends, and admirers
very anxious. They hoped that his return to his Motherland, to
his parents, to the friends of his early manhood, and to those who,
in the small Madras scientific world, admired his work — even if
unable to assimilate much of it — may have the effect of restoring
him completely to health and of the world being given further
instalments of the results of his wonderful genius. They heard
that somebody indeed wrote of him from Cambridge that since
Newton there has been nobody his equal — of which they felt
that there could be no higher eulogy.

J6 Last Months

Ramanujan spent about nine months in different places like
Madras, Kodumudi (a village on the banks of the Kaveri) and in
Kumbakonam — his home-town. The Government and the
University felt very concerned about the deterioration of his health.
In January 1920 he was persuaded to go to Madras for treatment.
Dr P S Chandrasekhara Ayyar, the Tuberculosis Specialist of
the Government of Madras, and other medical friends attended
on him. T Numberumal Chetty gave him his Chetpet House
free. S Srinivasa Ayyangar bore all his expenses, while some of
the Members of the Syndicate of the University of Madras also
made contributions towards his expenses in their individual
capacities. It was at this time that Ramanujan's mother went
to G V Narayanaswamy Ayyar-in great anxiety and pain as stated
in Sec B24. In spite of the best medical help available at that time

in Madras, Ramanujan died on 26 April 1920 at Chetpet, surrounded by parents, wife, brothers, friends, and admirers.

J7   Mathematical Meteor

Great men come and go like meteors. They shine and consume themselves prematurely. They blaze a new trail. Sankara was such an illustrious soul in philosophy, and Jesus in religion, so were Abel and Galois in Mathematics. These belonged to earlier periods. Ramanujan was such a one in our own times. He shone and consumed himself. His birth, his super-activity in Madras and Cambridge, and his death — all seemed to have happened in a flash.

CHAPTER K

MEMORIAL

K1 Portraits

The mathematicians and the University of Madras were filled with grief when Ramanujan passed away at the early age of thirty-three. One of the first memorials put up was an oil painting of Ramanujan. It was difficult to find any good picture of him in Madras. Therefore, the picture in his passport was used as the basis. I remember the portrait being unveiled by Sir K Srinivasa Ayyangar, the Vice-Chancellor of the University of Madras, in the Syndicate Room in 1922. A photograph of him in his academic dress as a B A of the University of Cambridge had already appeared in the *Journal* of the Indian Mathematical Society, vol 11;1919, Aug. A portrait based on the group photograph taken in Cambridge and in possession of Suryanarayana Sastry, Reader in Philosophy of the University of Madras, also adorns the walls of the University of Madras. Another copy of it was presented to Annamalai University by Dr A Narasinga Rao. Hardy gives a photograph of Ramanujan in his book of 1940 on Ramanujan. He says that he got the original for this from the Physicist, S Chandrasekharan F R S.

K2 Ramanujan Prize

In 1930, some of the admirers of Ramanujan belonging to the Indian Mathematical Society raised a fund of Rs 18,000 and instituted the Ramanujan Prize. It is administrated by the University of Madras. The prize is to be awarded on the basis of papers contributed for the purpose. I remember that S Chandrasekharan, then a student in Cambridge, and Kosambi, a student of the Harvard University, were the only two who competed for the prize in the first year. Chandrasekharan's paper was on Astrophysics, while Kosambi's was on Geometry. The Board of Assessors found it difficult to choose between the two papers. Therefore, they recommended that the prize be divided equally between the two competitors.

The following awards have been made in the later years:

Year	Name	Subject
1937	Dr S Sivasankara Narayana Pillai	Waring's Problems with Polynomial summands
	Dr B Ramamurti	Geometry of the Rational Norm Curve
1939	Not awarded	
1941	Dr T Venkatarayudu	Vibrations of a Symmetric Point System
	Dr S Minakshisundaram	A Problem of Fourier Expansion
1943	Not awarded	
1945	Dr B S Madhava Rao	Contributions to Algebra Related to Elementary Particles of Nature
	Sri V S Krishnan	Canonical Extensions of Partially Ordered Sets
1947 to 1958	Not awarded	
1959	Sri B R Srinivasan	On the Representation of Plane Conics by the Points of a Five Space

K3 Ramanujan Institute

K31 VIJAYARAGHAVAN AS FIRST PROFESSOR

Dr Alagappa Chettiar was a patron of higher learning. It was he who established the Alagappa Chettiar College of Technology as part of the University of Madras. He was too young to have known Ramanujan; and yet he established the Ramanujan Institute as a memorial to Ramanujan. He selected T Vijayaraghavan as the First Professor of the Institute (*See* Sec E62).

K32 INVITATION TO S S PILLAI

Early in 1950, one of Dr Alagappa Chettiar's advisers consulted me on the Institute. I recommended that the Institute would be more productive if there was another Professor also. For, I felt that research required communion among equals. I recommended the name of S Sivasankara Narayana Pillai (often referred to as S S Pillai), for this additional post. I was asked to negotiate

52

with him in this matter. A few days later, I had to go to Calcutta on some business. S S Pillai was then a lecturer in the University of Calcutta. I met him and took his consent. At the same time, I also persuaded him to accept the invitation of the Institute of Fundamental Research of Princeton, and spend one or two years there. He agreed to this also. He timed his flight to U S A so as to be in time for the International Congress of Mathematics being held in 1950 at the Harvard University. But, unfortunately the plane, in which he flew, crashed near Cairo and a most brilliant mathematician of India was lost. Pillai would have been a great asset to the Ramanujan Institute. He had qualities much in common with Ramanujan. In modesty and simplicity, he equalled Ramanujan. Like Ramanujan, he was a self-made man.

K33 ANECDOTE 6: S S PILLAI AND RAMANUJAN'S NOTEBOOK

S S Pillai had taken his B A degree in Mathematics in 1925 or so. But in those days the mathematics course did not include even Differential Calculus and yet he showed enormous insight into some of the abstruse problems in Mathematics. In 1925, I found him reading Mathew's *Theory of numbers* in the Madras University Library (*See* Sec P36). I thought that he was misled by the simple title of the book. I asked him whether he understood the theorems in it. When he said "Yes", I was taken by surprise. I took him into my room and showed him the "Frayed Notebook" of Ramanujan, which I had just then brought from England. While turning through the first chapter, he said that he too had worked out some Magic Squares. But he really got absorbed in the chapters in the Notebook dealing with Partition of Numbers. Thereafter, he was made a research student of the University of Madras. Between then and his death he had done a considerable amount of work on the Theory of Numbers.

K34 TRANSFER OF THE INSTITUTE TO UNIVERSITY OF MADRAS

Vijayaraghavan died on 20 April 1955. Thereafter, C T Rajagopalan became the Professor of the Ramanujan Institute. After the death of Alagappa Chettiar, the Ramanujan Institute had some financial difficulties. I remember discussing this with a high official of the Ministry of Finance of the Government of India. He said that the Government would gladly provide the necessary finance. Eventually, the Ramanujan Institute was transferred to the care of the University of Madras, with an existence independent of its Department of Mathematics.

K4 Postage Stamp

On 22 December 1962, the Seventy-fifth Birthday of Ramanujan, the Postal Department of India issued a special postage stamp with Ramanujan's portrait. Its denomination was 15 Naye Paise (*See* Sec A5).

K5 Ramanujan Exhibition

The University of Madras celebrated the Seventy-fifth Birthday of Ramanujan on 22 December 1962 (*See* Sec A5). In that connection, an exhibition of Ramanujan's Archives in possession of the University of Madras was held.

K6 Ramanujan Hall

During the celebration of the Seventy-fifth Birthday of Ramanujan by his old school — The Town High School, Kumbakonam, one of its main halls was named Ramanujan Hall.

CHAPTER L

RAMANUJAN'S NOTEBOOKS

L1 Acquisition by the University of Madras

By 1919, on the initiative of R Littlehailes, the University of Madras decided to establish a Mathematics Chair and invite Ramanujan to occupy it (*See* Sec H5). But his death in 1920 made this decision inoperative. However, the University acquired all the papers of Ramanujan including his Notebooks from his family and in return paid a monthly allowance of Rs 20 to Mrs Ramanujan. Later the University recognised the depreciation of the value of money after World War II and enhanced the allowance to about Rs 125 or so. Among the Ramanujan archives received were found two Notebooks — both well-bound. One was old and nearly filled up and the other was new and only a few pages of it were filled up. Reference has already been made to the latter Notebook in Section B5. While examining the archives for writing the biography (*See* Sec A2), I found a reference to another Notebook having been left behind by Ramanujan in Cambridge.

L2 Search for the "Missing Notebook"

The above information was found by me in August 1923, when I was still teaching Mathematics in the Presidency College. I could only guess that it should be the "frayed notebook" referred to by Ramachandra Rao and mentioned in Sec D3. At that time it did not occur to me that it should be searched for. But by an unusual coincidence of events, I became the first University Librarian of Madras on 4 January 1924. In fulfilment of one of the conditions of my appointment, I went to England in September that year and spent a year studying Library Science and observing libraries at work. I then decided that I should trace out the "frayed notebook" in Cambridge. But Hardy had gone to Oxford. Therefore, I first went to Oxford to get from Hardy some clue about the notebook (*See* Sec G5).

L3   Anecdote 7: Recovery of the "Missing Notebook"

It was March 1925. I met Hardy in the New College, Oxford.
He received me with kindness.

HARDY: Come in, Ranganathan. Edward B Ross has written
to me about you occasionally. The latest I heard from him was
that you have given up teaching mathematics and become the
University Librarian.

RANGANATHAN: Yes. I found a peculiar link between my
period as teacher of mathematics and my present period as librarian.

HARDY: What is it?

RANGANATHAN: While going through the Ramanujan archives
in the University to prepare a draft of Ramanujan's biography, I
found reference to a Notebook which he had left behind in Cambridge. This happened at the junction of my change-over from
teaching mathematics to caring for the library. A joint urge coming
from these two professions made me utilise my visit to England
in search of that Notebook of Ramanujan.

HARDY: You need not go a long way to search for it. (Then
Hardy went into another room and came out with the "frayed
notebook" in his hands). Here it is. Take it.

RANGANATHAN: Excuse me. If I had known that it was with you,
I would not have asked you about it. It is best kept by yourself.

HARDY: No. It is not right. Ramanujan belongs to your
country. The proper place for this Notebook is your own University
Library.

RANGANATHAN: I accept it, sir, with much reluctance and
almost with repentence.

HARDY: Do not have any such feeling. If you so desire, you
may send me a copy of this Notebook after you reach Madras.
It will enable several others and myself here to study the Notebook
and work on it.

RANGANATHAN: In that case, why don't you keep the Notebook
yourself and edit it for publication?

HARDY: That is too difficult a task. Actually I experimented
with working on one chapter of this Notebook—the chapter on
Hyper-Geometric Series. Here is the result. [(He handed over
to me a reprint entitled "Hardy (G H): A chapter from Ramanujan's Notebook" (Proceedings, Cambridge Philosophical Society.

21; 1923; 492-503)]. This work has taken me several weeks. If I am to edit the entire Notebook, it will take the whole of my life-time. I cannot do my own work. This would not be proper.

RANGANATHAN: I see your point. I shall certainly make a copy of the Notebook and send it to you.

HARDY: Also see if you can persuade your University to have it roughly edited, have some of the terms written in modern notation, and publish the Notebook for wider use.

L4 The Value of the Notebook

On reaching London with the Notebook, I wrote to the Madras University asking what I should do with it. The answer was very simple. It said, "Keep it with you". Evidently, there was nobody there to assess the value of this Notebook or become enthusiastic about it. Two weeks later Mr Esdaile, my teacher in the University College, London, and Superintendent of the Reading Room of the British Museum Library, showed me some of the treasures in that library. A quarto volume of Shakespeare was one of them. It was well-bound and kept in a very costly box. I jocularly remarked that the British were idol worshippers. "What do you mean?" asked Esdaile. I replied, "We in India do not make such a fetish of such an old book." This was rather strange to Esdaile because he was under the impression that we were the worst idol worshippers. To prove that we were not, I told him about the "frayed notebook" of Ramanujan, and the reply I got from the University about it. Esdaile told me, "Is that so? In that case why don't you hand over the manuscript to us? You can become rich by £500." I then replied, "Now that I know its value, I am not going to part with it!" We had a hearty laugh.

L5 Copies of Notebooks

The University allowed me the necessary funds to make three copies of each of the three Notebooks. The copying was generally done with the help of S Adinarayanan, an Honours Mathematics graduate and an old student of mine, Prof G A Srinivasan, and Dr R Vaidhyanathaswamy. One of the copies was sent to Hardy as desired by him. The other copies are preserved in the Madras University Library along with the originals. The copy that went to Hardy had been most productive. Several papers appeared in different learned periodicals in due course, based on that Notebook.   G N Watson worked through these Notebooks more systematically than others.. Between 1928 and 1936 he produced 25 papers based on these Notebooks. Hardy himself had produced about a dozen papers (*See* Sec Q7).

L6   Chronology of Notebooks

On my return to Madras, I deposited the "frayed notebook" by the side of the other two Notebooks already in the University Library.   There was no difficulty in deciding that the most fresh looking Notebook was the latest.   The matter in it was altogether of a different kind.   I began to compare the other two Notebooks. To my surprise their earlier chapters were almost identical.   As I proceeded further, the corresponding chapters looked like different versions of each other. Some of the chapters towards the end were totally different.

It was difficult to solve this puzzle.   At that time it happened that Ramachandra Rao was the Collector of Madras.   I called on him at his office, told him the story, and asked him if he could solve the puzzle.   He said that the "frayed notebook" was the first one.   He also added, "I asked Ramanujan to make ā copy of that Notebook for the sake of safety.   But it is so difficult for a creative thinker like Ramanujan to make a faithful copy as clerks of the Sub-Registrar's office are expected to do.   Evidently he began with the idea of making a faithful copy.   Later on, he went on improving the chapter while copying them.   Finally he began to enter altogether new results in new chapters."

L7   Delay in Publishing the Notebooks

I informed the University Authorities about the second wish of Hardy that the Notebook should be printed after moderate editing.   But for some reason or other it was not carried out. The wish of the Board of Studies in the matter was not heeded to. After my joining the University of Delhi in 1947, I had occasion to mention about this Notebook to Sir Maurice Gwyer, Vice-Chancellor of the University and former Chief Justice of India. He said, "It is a shame that the Notebook has been left unpublished over twenty years.   If you get the Notebook, I shall get it published."

58

L71   HOPE FOR PUBLICATION
The Commonwealth Universities' Congress was held in Oxford
in 1948.   The Vice-Chancellors of practically all the Indian Universities were present.   The Delhi University was represented by me.
Dr Sir S Radhakrishnan, Dr Wali Muhammad, the Vice-Chancellors of about three Indian Universities, and myself were sitting
together in a hall one afternoon. Before the function of the
occasion began, our conversation somehow landed itself on
Ramanujan's Notebooks. In answer to a question by Dr Radhakrishnan, I mentioned Sir Maurice's offer to have the Notebooks
published by the University of Delhi. Dr Radhakrishnan asked
me how he could find money for it. I replied that Sir Maurice
had said that he would send his hat round if need be. Dr Wali
Muhammad asked what the cost of the publication would be. I said
that it might not exceed Rs 10,000. On hearing this amount,
Dr Sir A Lakshmanaswamy Mudaliar said that it was not worthwhile to send the hat round for this trivial sum and that the
University of Madras itself could meet the cost. As the manuscripts
were the property of that University, everybody felt that the
publication of the Notebooks by that University would be most
appropriate.

L72   PURSUIT OF THE HOPE
A few days later, another event occurred in London. I was
living in the Kensington Hotel at the Palace Gate Road.   One
morning, a Chinese gentleman called at my room.   I now forget
his name. He said that he saw my name on the hotel's name-board
and that led him to look for me. He added that he was a mathematician who had worked for some years in Cambridge and was
on the way to the United States of America to work as a Professor
in some University. He continued that he had heard my name mentioned in Cambridge by somebody in connection with Ramanujan.
As he himself had been working on some of Ramanujan's results,
he said that he was keen to meet me. He showed a knowledge
of the copy of Ramanujan's Notebooks circulating in Cambridge.
He asked why the University of Madras was not publishing it.
I referred to the conversation in Oxford and said that we were
on the eve of doing so. I asked him whether he could help me in
finding out a mathematician who would be able to edit the Note-

books providing proofs wherever possible. He said that it would be difficult for one person to edit all the chapters. Then he gave me a list of a dozen names indicating against each the chapter of the Notebooks that could be entrusted to him. It also struck me that Dr K Chandrasekaran, then working in the Fundamental Research Institute at Princeton and now Professor of Mathematics, Eido Technische Hochschule, Zurich, who was a young mathematician with a considerable drive, would be able to co-ordinate the work of the various editors and see the publication through the press. A few days later I happened to meet Dr A Lakshmanaswamy Mudaliar while walking along some street in London. I told him about this proposal. He approved of the idea and asked me to write to him about it after both of us would have reached India. I wrote accordingly. I suggested that photostat copies might be made of the various pages and that a copy of the relevant chapter might be sent to the editor concerned. In 1949, the University of Madras made three photostat copies of each of the three Notebooks. One of these was given to S S Pillai and it was lost in the air disaster at Cairo where he himself lost his life. The other two sets are preserved in the Madras University Library.

L8 Another Attempt

The Indian Mathematical Society met in Delhi in 1954. I then tabled a resolution that arrangement should be made to have the Ramanujan's Notebooks edited and published. Dr Sir K S Krishnan intervened and said that a resolution of that kind would be of no use and that a more effective way would be to make the Prime Minister interested in the matter. I agreed with this idea. Evidently at the suggestion of Dr K Chandrasekaran of the Tata Institute of Fundamental Research, and of Dr H J Bhabha, the Director of that Institute, the Tata Trust brought out a photostat edition of the three Notebooks in 1957. It is bound in two volumes. A thousand copies were said to have been issued for the benefit of the research workers and institutions, in the absence of a fully edited printed edition.

REMINISCENCES OF FRIENDS

MA   Introduction
This chapter contains the reminiscences of some of the surviving friends of Ramanujan and of Mrs Ramanujan. They are in five groups: school friends, college friends, Cambridge friends, other friends, and wife. Each Reminiscence is headed by the name and some details of the person concerned.

MB/MC SCHOOL FRIENDS
MB   Reminiscences of Mr N Raghunathan
Schoolmate of Ramanujan and retired Professor of Mathematics.

MB1   EARLY PROFICIENCY IN MATHEMATICS
Ramanujan and myself were both ¡in the same class and section in the Second Form in the Town High School, Kumbakonam, in 1899. I used to hear about him even in the preceding year as a mathematical genius; but he was in another section. We generally sat side by side in the Second Form. Most of our classmates used to go to him for help in Mathematics. Euclid was taught in English in the Second Form. It was a very formal subject and none of us — except Ramanujan — really understood the axioms, postulates, and definitions. Ramanujan used to help us in this subject. Next year, we got separated. I came back to Kumbakonam only in 1906 for my B A. But in spite of my absence from Kumbakonam, our acquaintance was not cut off at all, though we were not intimate except for a few months in 1899. During 1906-07, I used to meet Ramanujan occasionally in the Government College when he came to take out some book from the college library or see the lecturers. We used to talk of the "old days" on these occasions. Ramanujan was such a simple soul that one could never be unfriendly towards him. He had always a contemplative look and somehow I could not imagine anyone getting very intimate with him. Of course, none of us — in those days or even later — could really understand the nature of his genius. ·We used to admire his skill in solving

61

complicated questions in Arithmetic and Algebra and in solving geometrical riders; there our appreciation stopped. To be frank, neither Prof Seshu Ayyar nor Patrachariar nor any of the mathematical higher-ups in India correctly gauged the extent and nature of his capacity.

MB2 UNEASY IN EUROPEAN DRESS

I met him again only in 1914 after I joined the staff of the Presidency College. He was then employed in the Madras Port Trust. I went to his house once or twice. I was with him for the last time when he came to the college just before leaving for England, dressed in European style and feeling very uneasy in that dress. I heard about his return to Madras later, but I could not meet him before he died. A few years later, Seshu Ayyar sent for me and asked me to arrange for the cremation of Ramanujan's father who had come to Madras and suddenly died in Narayana Ayyar's house.

MC Reminiscences of Mr N Govindarajan

Retired Chief Engineer, Madras, and Member, Union Public Service Commission, New Delhi.

MC1 EARLY ABSORPTION IN MATHEMATICS

Ramanujan was a neighbour of mine in the years prior to 1911 and lived about ten doors to the west of my house in Sarangapani Sannidhi Street, Kumbakonam. Almost every house in the South (particularly in Kumbakonam) had a two-tier *pial* — one about two feet above street level and the other two feet higher still. Almost every morning, Ramanujan used to sit in the upper *pial* and he would be surrounded by his schoolmates and others who turned to him to solve their mathematical problems. Ramanujan would sit right on the *pial* floor with his legs folded vertically up and very many of the schoolmates used to amuse themselves by filling his underwear with small pebbles. That Ramanujan would not notice it, was a measure of the extent to which he would lose contact with the world without, when under the spell of his beloved science.

MC2 DRAFTING SCHOOL TIME TABLE

In the Town High School, Kumbakonam, having perhaps 1,500

pupils on its rolls and staff of 30 or more teachers, the preparation of the time-table without internal conflict was a problem. It was the duty of Ganapathy Subbier, the senior mathematics teacher of the school to prepare the time-table. He had so much confidence in the analytical and synthetic ability of Ramanujan, that year after year, he entrusted the preparation of the time-table to our Ramanujan.

MC3 ABANDONED AS TUTOR
When I passed my F A in 1908, Ramanujan was a free lance, working on his problems. I requested Ramanujan to give me tuition in B A mathematics—particularly in Differential Calculus. He was my tutor for a few days—say a fortnight; but in almost all those days, he would talk only of infinity and infinitesimal and I felt that his tuition might not be of real use to me in the examination and so I gave it up—certainly not the first instance of the world finding one of its geniuses unfit for its day-to-day commercial necessities!

MC4 PROPHECY ABOUT FUTURE GREATNESS
My father developed astrology on somewhat rational lines. After examining Ramanujan's horoscope, he predicted that, his horoscope indicated his becoming a great scholar about his 32nd year. This he said in 1912, when the greatest mathematical prodigy of this century was buried in the labyrinths of the Port Trust Office as an obscure clerk.

MD/MK COLLEGE FRIENDS
MD Reminiscences of Mr K Chengalvarayan
Advocate, Chingleput

MD1 SOCIABILITY AND INNOCENCE
Ramanujan was my classmate in the Senior F A Class in Pachaiyappa College, Madras. From what I know of him personally, I can say that he was a sociable person. Innocence and absence of self-consciousness were his other qualities. He was unconscious of his own abilities and attainments. I may not be wrong when I say that other eminent mathematicians discovered Ramanujan for him. Early in the term after we joined the Class,

I had many doubts and difficulties in the course of my work. I wanted to take help from fellow students, now and then. But, I felt it rather humiliating to expose my backwardness to students very good at Mathematics, of whom there were a few in the class. I wanted, therefore, to consult an ordinary student only—just above me in ability. One afternoon in the courtyard of the College, I casually asked Ramanujan whether he had finished working the chapter in Radhakrishna Ayyar's *Algebra* assigned for home work.

MD2   PUZZLE TO ME

The reply to my question came from C R Krishnaswami who exclaimed, "What! You are putting that question to Ramanujan!" I took the reply to mean that Ramanujan finished working that chapter and other chapters too. Soon after that incident, Ramanujan became a puzzle to me when some students of mathematics in the B A class sometimes sought his help on the subjects, and when I saw and heard how he was regarded by N Ramanujachari and other Professors of our own and other Colleges.

MD3   LIFE IN ENGLAND

Subsequent career of Ramanujan is known to me only from conversation among students. Whenever I met my classmates — that was at long intervals — the emergence of Ramanujan into fame was usually one of the topics of talk. In the course of such talks, I heard that for some time Ramanujan stayed with Suryanarayanan in England and that both cooked their food together. Suryanarayanan was a brilliant student of Philosophy.

ME   Reminiscences of Mr T Devaraja Mudaliar
Indian Police Service (Retired), Madras

ME1   MODESTY

The late S Ramanujan joined the senior F A Class in the Pachaiyappa College in 1907, after having once failed in the F A Examination. Mr Radhakrishnan, Advocate of Tanjore, and I were among his classmates. Ramanujan was very modest. He was somewhat weak in English and subjects like Physiology. He could have easily become proficient in these subjects. But on account of his devotion to Mathematics, he could not find time to study them.

ME2 APPRECIATION BY N RAMANUJACHARI

N Ramanujachari was teaching Mathematics to the F A Class. He would take the whole of two sliding blackboards to solve a problem in Algebra or Trigonometry and arrive at the answer in the tenth or eleventh step. It was difficult for average students to follow him until he explained each step twice or thrice. Then, Ramanujan would stand up with diffidence and ask for his permission to solve the problem in a shorter way. After the Professor gave the permission, he would arrive at the answer in the fourth step. Ramanujachari was somewhat deaf. He had the habit of bending his head towards the speaker to catch his words. When Ramanujan was mentioning step after step, Ramanujachari would repeat, "What? What?" Only after Ramanujan quoted the formula which he used and explained it twice or thrice, he could follow him. It was all Greek and Latin to the others.

ME3 APPRECIATION BY P SINGARAVELU MUDALIAR

The Chief Professor of Mathematics was P Singaravelu Mudaliar. He was teaching only students of the B A Classes. He was an acquisition for Pachaiyappa College. He was formerly Assistant Professor of Mathematics in the Presidency College and was the first to open the Mathematics Department in the Pachaiyappa College. His reputation was widespread as a very successful Professor. Ramanujan usually brought his tiffin in a brass box — evidently curd and rice with pickles — not sumptuous food at all. After hastily finishing his meal, he would go to Singaravelu Mudaliar who waited for Ramanujan's assistance to solve difficult problems in the current Mathematical Journal. They would jointly solve them.

ME4 PHYSICAL APPEARANCE

Ramanujan was of a somewhat stout built; but he looked pale and anaemic. If I were one of his intimate friends, I would have advised him to take two raw eggs every morning. But he would have rejected my advice. For he was orthodox in his social practices. While in England, he had an attack of tuberculosis and was admitted into the hospital. He begged his Physicians to give him any medicine they thought fit, but not animal food.

ME5 LEARNING IN RELIGIOUS LORE

Mr Radhakrishnan, one of our classmates and Ramanujan's intimate friends, told me that on a moonlit night, both walked from Kumbakonam to Nachiarkoil to witness a religious festival. On their way, Ramanujan recited passages from the Vedas and the Sastras and explained their meaning to him in a very instructive manner.

MF Reminiscences of Mr C R Krishnaswami Ayyar, Pleader, Tirupattur

MF1 SHARING OF SCHOLARSHIP

I was a classmate of the late Srinivasa Ramanujan in the Junior F A Class in the Pachaiyappa College in 1906. A few days later, I approached the Principal of the Pachaiyappa College, J A Yates, for a full free scholarship in the Junior F A Class. He granted me my request and asked me to come the next day with only the library fee of half a rupee. The next morning, I went to the Principal as directed. Ramanujan had just then been introduced to him by Prof N Ramanujachari, who was teaching mathematics for the F A Class. He appears to have told the Principal that Ramanujan had a special aptitude for mathematics and that he (the Professor) himself had found some of the mathematical problems and theorems which Ramanujan had worked out in his Notebook to be very startling and difficult to follow. Thereupon, the Principal cut the full free scholarship he had given to me into two halves and made us accept one half each.

MF2 APPRECIATION BY N RAMANUJACHARIAR

During the mathematics classes, Ramanujan would be working his own mathematical problems in a corner of the top gallery. When Prof Ramanujachariar sometimes drew Ramanujan's attention to a problem he was working out on the blackboard and asked him to give the next step or continue the problem, Ramanujan would have the boldness to say that the several intermediate steps worked out by the Professor were unnecessary and he would jump to the last step and give the answer quite to the surprise of the whole class and much more of the Professor. The Professor would then ask him to come down to the blackboard to explain to the class how he had arrived at the answer.

MF3 TOO POOR TO BUY A CAP

Ramanujan and I had Sanskrit for our second language. One day he came to the Sanskrit class in the morning wearing his usual coat of black Calicut check, but without his usual wool-knit red Hassan cap, which, he said, had been blown off by a gust of wind when he got into the tram. Our Sanskrit Pandit, Krishna Sastriar (the father of Patanjali Sastriar, retired Chief Justice of India) asked him to go out and get himself a cap from the adjacent China Bazaar shops, as attending class without a head-dress covering his tuft was against discipline. Ramanujan begged to be excused saying that he was too poor to buy another cap, though its cost was less than half a rupee.

MF4 WATER POURED OVER HEAD

In 1912, Ramanujan shared an upstairs room in Sami Pillai Street, Triplicane, with me (then a student in the Law College) and a cousin of mine who was a clerk in the South Indian Chamber of Commerce. One night, when Ramanujan was speaking to me about the wonders of the astronomical world, my cousin, who felt his sleep disturbed, poured a pot of water over Ramanujan's head, saying it would cool down his heated brain. With the patience of a mathematician, Ramanujan exclaimed with joy that he had a good "Gangasnanam" (bath in the River Ganges) and would like to have another downpour!

MF5 EVALUATION BY P SINGARAVELU MUDALIAR

Ramanujan used to work some mathematical problems and show them to the Chief Professor of Mathematics, Singaravelu Mudaliar, who was taking Mathematics only for the B A Classes. He used to take home the problems or certain queries of Ramanujan and return them to him the next day, frankly confessing that he was not able to find the solution and would still further work at them. He was a mystery to him and the Junior Professor, N Ramanujachari.

MF6   AVERSION TO PHYSIOLOGY

We were obliged to study Physiology as a compulsory subject for the F A Examination. Many students used to be ploughed at the F A Examination (not even 20 per cent passes on the whole) owing to the rigour of the valuation in such an unimportant subject then. For, the Examiner was a European Chief Professor of Physiology in the Madras Christian College. Our textbook of Physiology was by Forster and Shore.

MF61   SEA-FROG AND WELL-FROG

Our Professor of Physiology was B G Duraiswamy Pillai. He used to bring a number of sea-frogs — very big ones — to the dissection table. While the students dissected them, he used to explain the different parts of the structure of the frog. He would emphasize that the structure of the frog was nearer to that of man. Ramanujan and myself abhorred the dissection work. After administering chloroform to the frog, suprisingly, Ramanujan asked the Professor, "Sir, are the sea-frogs chosen for dissection because we are all well-frogs?" Thereby, he insinuated the smallness of Man as against the marvels of the magnitude of the Universe. The Professor smiled and carried on his further instructions about the frog structure. This reminds me now of the conversation between a sea-frog and well-frog depicted by Swami Vivekananda in one of his lectures.

MF62   SERPENT THEORY IN YOGA

Then again, Ramanujan pressed another query on the Professor. "Sir, where is the Serpent in this frog structure (if you say it is like that of a man), which would swallow the frog itself?". Ramanujan evidently meant the Serpent Power or *Nadi* in man — the *Kundalini Sakthi*, with six centres one above another starting from the base of the vermiform appendix and proceeding upwards to the crest of the human structure. Ramanujan was familiar with the tradition of Patanjali's *Yogasutra*, later rendered into English by Arthur Avalon (pseudonym of John Woodruffe.)

MF63   Indigestible Digestion

A third instance in connection with the physiology class is this. Ramanujan did not take real interest in that subject nor attach any importance to Forster and Shore's *Physiology* or any other book on that subject. Therefore, he did not fare well in that subject. The Professor of Physiology used to have one examination for the class after he finished one chapter in physiology. Once the examination paper had a question on the Digestive System. Ramanujan, after receiving the question paper, wrote a few lines on digestion and closed his answer, with the words, "Sir, this is my undigested product of the Digestion Chapter. Please excuse me". He scrupulously avoided putting his signature to the paper, though the Professor rightly suspected that it would be Ramanujan's and asked whether it was his answer book. Ramanujan confessed his 'guilt' with silent humility.

MF7 GOVERNOR'S TRIBUTE
In 1915, when presiding over a meeting of the Presidency College Mathematical Association, Lord Pantland, the Governor of Madras, said, "The Madras Presidency is very lucky in producing a mathematical genius like Ramanujan. England wanted the valuable services of such a genius in mathematics as Ramanujan for some more years. He cannot be sent back to India so soon."

MF8 ON SEEING RAMANUJAN'S PORTRAIT
When I visited the University Library in 1937, I saw the portrait of my friend Ramanujan on the walls of the Reading Room. Tears flowed from my eyes involuntarily.

MG Reminiscences of Mr K Narasimha Ayyangar, Advocate, Kumbakonam.

MG1 COACHING BY RAMANUJAN
In 1911, Ramanujan lived with me in 9 Venkatarayan lane, P R Square, Park Town, Madras. My brother and myself lived with him like brothers. He was a loving and kind person. For instance, during that period, I had the good fortune of being coached by Ramanujan in Mathematics for my Intermediate Examination. On the day of my Mathematics Examination, he instinctively felt that he should meet me and persuade me to sit for the examination in the afternoon also without fail. He, therefore, came all the way from our house walking to the Presidency

College, a distance of four miles. He found me depressed and hesitating whether to stay away from the afternoon examination. He encouraged me and gave me some last minute tips as if by prophecy or intuition and persuaded me to attend the examination, and I found his prophetic tips very useful; and they must have contributed to my passing the examination.

MG2   INTEREST IN TRADITIONAL LORE

Incidentally it should be mentioned that Ramanujan knew and was interested in Astrology, Vedanta Philosophy, and various theories like the theory of relativity. His discourses on them were astounding and entertaining.

MG3   DECLINES INVITATION TO CAMBRIDGE

When Ramanujan was living with me, he was unemployed and was spending most of his time in our house with his mathematical research. During that period I had occasion to take him to several gentlemen interested in higher mathematics including my own Professor Edward B Ross. In 1912, he lived separately with his family in Triplicane. As I too moved to Triplicane by that time, we were in constant touch with each other. Some of us were keen that he should be introduced to the mathematicians in Cambridge. We persuaded him, therefore, to send some of his results to Professor Hardy. This led to Hardy asking Ramanujan whether he would go to Cambridge as a research scholar. But without telling any of us, he appears to have declined the offer.

MG4   ACCEPTS OVERSEAS SCHOLARSHIP

In 1914, the University of Madras gave him an Overseas Scholarship. This time he told us about this offer and asked for our advice. We wrote out a reply accepting the offer, got his signature in it, and posted it.

MG5   VOYAGE TO ENGLAND

Sir Francis Spring, Chairman of the Madras Port Trust and his one-time master and benefactor, insisted that Ramanujan should embark only in the Madras harbour and Professor Neville, who had come to Madras from Cambridge to deliver lectures in mathematics in the University of Madras, kindly agreed himself

70

to take the boat at Madras instead of Bombay. Doctor Muthu, a specialist in Tuberculosis, was also travelling by the same boat; with the result that Ramanujan, very little conversant with worldly affairs and having very little capacity to move freely with others, or to take care of himself, had two friends to take care of him during the voyage and on arrival in England. The day previous to his sailing, my cousin Aravamuda Ayyangar and myself had the privilege of having him in our room in Triplicane and the whole night we spent talking about men and things and the prospects of his stay in England and the next morning we saw him off at 9 a m in the Boat. Even during the voyage, he seems to have improved his English. For, the two letters he wrote to me during the voyage were in good English. Unfortunately, they have not been preserved.

MG6 EXPERIENCE IN CAMBRIDGE

After reaching Cambridge, Ramanujan wrote to me that the first few days were not very congenial, in that he was not communicative enough with the other students in the University. This caused some little mutual annoyance. In a short time, however, the students began to treat him with great respect.

MG7 THE LAST MEETING

When he arrived at the Madras Central Station, I met him on the platform. But, alas! I found him not as the cheerful, chummy, and affectionate Ramanujan that he was — specially with me — but a thoroughly depressed, sullen, and cold Ramanujan — even on seeing me a close and affectionate friend of his. When he came again to Madras after some time for treatment, I met him at his residence in Harrington Road, Chetpet, and found to my great grief that though physically living, he appeared to be preparing to leave. Soon after, his tragic end came; and thus ended prematurely a career, which was bound to have brought immense further glory to himself and to his country too.

MH Reminiscences of Sri M Patanjali Sastry
Retired Chief Justice of India

MH1 SIGNIFICANCE OF SEVEN

I knew Ramanujan as a junior contemporary of mine in

Pachaiyappa College; but I did not have many contacts with him. He was a classmate of my last cousin, M Ramaswami Sastry, later Mathematics Assistant in the Pachaiyappa High School. I remember Ramanujan once explaining to us the significance of the number "7" in the theory of numbers as understood by our ancestors — who spoke of Sapta Rishis, Sapta Dweepas, Ratha Saptami, etc—much of which I was unable to follow. I wonder if there is any reference to this in his famous Notebook.

MH2 TALENT FOR MATHEMATICS

I remember Professor Singaravelu (then Professor of Mathematics in Pachaiyappa College) once telling me that he was unable to understand some of Ramanujan's equations and that the young man's talent for Mathematics was extraordinary.

MJ Reminiscences of Mr R Radhakrishna Ayyar, Advocate, Tanjore

MJ1 TAKEN TO BE AN ORDINARY STUDENT

I knew Ramanujan through one of the few lucky accidents of life. I was a student in Pachaiyappa College in 1906-1907, when I was doing my F A. Boys were seated in an alphabetical sequence. Rāmanujan sat next to me or was probably the second from me. Since we were together in that fashion for quite a long time, our acquaintance grew and ultimately blossomed into friendship. If I could not have foreseen that I was in company with a prodigy whom the world recognised, the misfortune was mine. But I may as well take solace in the fact that none at that time, barring a few, knew of the coming of a genius.

MJ2 PHYSICAL APPEARANCE

Ramanujan was fair and plumpy and his build reminded one of a woman since his palms were smooth having nothing of the rough characteristics of a man. When he thought hard, the pupils in the eyes would vanish making it appear that he had a squint or something like that. I do not recollect his taking to any games worth mentioning.

MJ3 POVERTY

The prodigy whom Kumbakonam gave to the nation came of

72

a poor family. When I say this, I am not saying anything from guess. In those days, it was necessary for college boys to wear buttoned-up coats and some head-gear. Ramanujan invariably wore a buttoned-up check coat which hid no shirt within. A Hassan cap was almost characteristic of him. He lived in a small lane off the Fruit Bazaar in Broadway. The unpretentious room which was his "home", run by his grandmother, spoke of his station in life. It was not only dark and dingy but also ill-ventilated. I remember visiting him in his room occasionally, though we met mostly in the college.

MJ4 SOCIABLE QUALITY
I want to dismiss a popular impression that Ramanujan was either morose, or unsociable. He enjoyed swapping funny anecdotes and his own way of narrating them was to reveal something and give out a hearty laugh with a guffaw which he usually did by cupping the mouth with a palm.

MJ5 HELP TO PROFESSOR
Did Ramanujan show any signs of the coming of the prodigy? This is a question which will naturally be asked of one who claims to have known him rather closely. N Ramanujachari was the mathematics lecturer. Whenever he had some problem while lecturing, he would turn to Ramanujan and ask him: "I say, what do you think about this?"

MJ6 GOD NARASIMHA'S DIRECTION
I had known Ramanujan talking to Mr P Singaravelu Mudaliar, Professor of Mathematics, and placing before him the papers relating to his research. When I visited him thereafter, I asked him about his research. He said that Lord Narasimha had appeared to him in a dream and told him that the time had not come for making public the fruits of his research and that he must wait for the opportunity which would come. It would thus be obvious that Lord Narasimha was his favourite Deity. Ramanujan was immensely devout.

MJ7 KNOWLEDGE OF ANCIENT LORE
Many a time, I have had the good fortune of listening to his

73

philosophical dissertation on the Ramayana and the Mahabharata in which he was very lucid.

MJ8 OBSTINATE PATIENT

In 1909, I passed my B A examination and was doing my F L. I was living in Sundaramurti Vinayagar Koil Street, Triplicane. Some friends ran a mess on a co-operative basis. It was at this time that Ramanujan, who was living in George Town, became seriously ill. His landlord wanted him to go and live with some friends and so it chanced that Ramanujan came in a horse-cart to my place as I was starting for the college. I was moved to see him in bad shape and took him out of the cart and put him in my bed. Our cook was asked to attend on him. As a patient, Ramanujan was not exemplary; he was obstinate and would not drink hot water and insisted on eating grapes which were sour and were bad for him. I called in Dr Narayanaswami, who after examining him, asked me to send him to his parents as his condition required constant nursing.

MJ91 CONCERN ABOUT HIS NOTEBOOKS

I remember taking Ramanujan to Beach Railway Station and putting him in charge of the Superintendent of R M S, who was my friend. Before going to Kumbakonam, Ramanujan gave me two big notebooks which, I little knew, contained the outpourings of a genius. They were closely written. I cannot recall what he told me without suppressing a tear. He said they were for safe keeping. "If I die, please hand them over to Prof Singaravelu Mudaliar or to the British Professor — Edward B Ross — of the Madras Christian College", he said. I thank God that he lived to return and take the Notebooks back from me.

MJ92 INTRODUCTION TO R RAMACHANDRA RAO

The other thing that stands out in my memory is the little help I was able to give him in 1911 when he was fighting the battle of life. I wrote to my father-in-law, Mr Gopala Ayyar, B A, B E, who was District Board Engineer at Nellore and requested him to help Ramanujan. I knew that the then Collector of Nellore was Diwan Bahadur Ramachandra Rao, who was keen about Mathematics. Later, I learnt that Mr Rao frankly confessed that he was out of touch with Mathematics but promised to help him by a monthly payment of Rs 20 or Rs 25 which he had said Ramanujan could collect from his (the Collector's) family in Mylapore. No doubt this helped him for some time.

MJ93   THE LAST MEETING

I did not see him before he left for England. I had heard that he had cried when he had to clip his abundant tuft. Fate had ordained that I should see him almost in his death bed in Kumbakonam. He was down with T B and in very poor shape. His brother told me that he was not in a position to see anyone. But when he did allow me in, I stood near a ramshackle cot in which the wizard, who today receives the homage of the world of mathematics, lay as a bundle of bones. Ramanujan had powerful and penetrating eyes. For a moment his eyes flashed a look of recognition and he mumbled almost inaudibly: "Radhakrishnan?" I am happy that I had the *darshan* of a friend I had known during a relatively brief spell of life and it was given to me to have known him and enjoyed the warmth of his company and friendship.

MK   Reminiscences of Mr Ramayana Ratnakara T Srinivasa Raghavacharya

MK1   PHYSICAL APPEARANCE

Ramanujan was my classmate in the Senior F A Class in 1906 in the Pachaiyappa College, Madras. I was seated almost next to him. He looked more like the late Dr Sir K S Krishnan. I do not approve of his picture in the postal stamp. During those years he was hale and healthy. To a certain extent he was uncouth in his dress probably due to poverty.

MK2   MENTAL ABSORPTION IN MATHEMATICS

He was often absent-minded, heavily engrossed in mathematical research, which made many think that he was brain-sick. Of the problems given in our textbooks in Geometry, Algebra, and Trigonometry he used to remark, "These are all mental sums". He rarely got more than 10 per cent in Physiology for which subject he had a supreme contempt and got something more, say 15 per cent to 20 per cent in Greek and Roman History, but managed to get about 25 per cent in English. He used to tell me, "I will give you shortly some startling theory of numbers." But he never told me anything about it.

MK3 COLLECTING PACKING PAPER FOR WRITING

So much about my college experience with him. About a few years later I accidentally met him near the Madras Harbour, when he was employed in the Port Trust. He was picking pieces of packing paper.

SRINIVASA RAGHAVACHARYA: What are you doing?

RAMANUJAN: I am picking up pieces of plain unwritten paper used for packing. For, I am in want of paper to write mathematical problems.

Sometimes he used to write notes and problems in red ink across already written paper for want of plain paper.

ML/MN CAMBRIDGE FRIENDS

ML Reminiscences of Mr K Ananda Rao, Retired Professor of Mathematics, Presidency College, Madras. (With acknowledgement to Number Friends Society, Muthialpet High School, Madras).

ML1 HELP OF R KRISHNA RAO

There was a person, not so well known, who stood by Ramanujan during his difficult days and was of considerable help to him. This was R Krishna Rao who was a cousin of my mother. I know the extent to which Krishna Rao rendered help to Ramanujan. Krishna Rao's association with Ramanujan deserves to be better known.

ML2 LIFE IN CAMBRIDGE

Ramanujan went to England a few months before I did, and returned to India a few months after my return. At Cambridge, he was at Trinity College and I was at King's. I used to meet him often during this period. During vacations, he used to stay mostly in Cambridge. His room was electrified and was provided with a gas stove. He cooked his food and his menu generally consisted of cooked rice, curds, fruits, sambhar with potatoes and other vegetables.

76

ML3 SOCIABILITY

By nature he was simple, entirely free from affectation, with no trace whatever of his being self-conscious of his abilities. He was a man full of humour and a good conversationalist, and it was always interesting to listen to him. He was very popular with the Indian students at Cambridge and he had also many English friends with whom he moved freely. He was often seen at tea-parties and other social gatherings, where he, of course, rigidly adhered to his vegetarian habits.

ML4 AFFECTION FOR MOTHER

He rarely talked of his family except for occasional references to his mother for whom he had great affection. He never mentioned to me even once about his brothers.

ML5 RANGE OF INTERESTS

On the occasions when I met him, we used to talk in homely Tamil. He could talk on many things besides mathematics, and he was particularly interested in psychic research and the experiments of Oliver Lodge and others.

ML6 UNASSUMING NATURE

I recall an incident which showed his unassuming nature. When I told him that I was preparing an essay for Smith's Prize, he went and asked Hardy in dead earnest whether he could also try for that Prize. Hardy had a hearty laugh over his question and assured him that what he had achieved deserved a far higher recognition than a Smith's Prize.

ML7 THE DEPTH OF RAMANUJAN'S CONJECTURES

His best work bears the stamp of remarkable originality. It is still a mystery how he was able to arrive at his conjectures of some very deep results. The unique underlying feature of Ramanujan's work was often of a staggering character. In this ability, Hardy compares Ramanujan to Euler and Jacobi. Ramanujan's example may give the hope that the cultivation and exercise of such skill have not yet reached their finality. A close study of his writings is worthwhile and may provide clues to his techniques. I cannot escape feeling that if Ramanujan had lived longer, he could have obtained far greater and more wonderful results in mathematics.

MM Reminiscences of Dr C D Deshmukh, Former Finance
Minister of Government of India, Former Chairman of the
University Grants Commission (India), and Vice-Chancellor
of Delhi University.

MM1 THRILL IN BEING A CONTEMPORARY IN CAMBRIDGE

To my generation, who matriculated in the early years of the
second decade of this century, the romance of the find of
Ramanujan and his being enabled to proceed to Cambridge for
advanced studies in Mathematics were well known from the newspapers.   Nevertheless, it was a thrill to me to discover on reaching
Cambridge in July 1915 that I was going to be a contemporary of
Ramanujan at this famous seat of learning.   Indeed, but for the
accident of my having earlier been admitted to the Jesus College
through the good offices of Prof Müller, who taught History at
the Elphinstone College, Bombay, I might have been at the same
college at which Ramanujan studied, that is, Trinity.

MM2 LIFE IN CAMBRIDGE

In those days Cambridge was a ghost of its normal self because
of the World War; and the much smaller number of students in
its different colleges found it easier to get to know one another.
This was especially true of the Indians then up at Cambridge.
It was not long, therefore, before I got to know Ramanujan.
Our acquaintance developed fast into close friendship. Ramanujan's current achievements were common talk amongst us Indian
students.

MM3 RAMANUJAN: HIS OWN COOK

Occasionally I used to have a Sunday meal with him at which
he apologised in a solicitous way about the spicy *rasam* that he
had prepared for his friends.   It is necessary to explain here that
Ramanujan was a very strict vegetarian.   Because he suspected
that even vegetarian food ordered from the college kitchen would
be polluted with undetectable animal fats, he used to cook food
for himself and his guests in person.   The meals so served were
quite delicious, especially as Ramanujan used to regale us with

mathematical conundrums and puzzles which he took care to ensure were not above the heads of his non-mathematical friends.

MM4 RAMANUJAN'S STRICTNESS IN VEGETARIANISM

It so happened that Ramanujan selected the same lodging house in London in later years that I had selected. We knew through common friends about it and the charming lady who ran this lodging house, which, incidentally, had had the honour of counting amongst its lodgers a few years earlier Vithalbhai Patel and Vallabhbhai Patel. The landlady told us of an incident which happened during one of Ramanujan's visits to this place in early 1918 when the Zeppelin raids over London were at their worst. Ramanujan had come to the place on a casual visit and had taken some breakfast which he was assured was purely vegetarian. Later, however, Ramanujan was told that the beverage served to him was ovaltine and Ramanujan, in idle curiosity, looked at the tin. He was horrified to find from the legend on the tin that the contents included a little powdered egg. This so upset him that he immediately packed his bags and left the place for Liverpool station on his way back to Cambridge. Later, the landlady received a letter from him to inform her that as he approached the station, there was heavy bombing of the neighbourhood and that it was with great difficulty that he could get a train for Cambridge and get away from London. He had added that he was convinced that the raid was a punishment meted out to him by God for having partaken of anything non-vegetarian.

MM5 RAMANUJAN'S CHILD-LIKE SIMPLICITY

This incident illustrates very vividly the child-like simplicity which was such an engaging trait of Ramanujan. His was a most lovable personality with a God-fearing temperament and a humility which is seldom found in such geniuses.

MM6 RAMANUJAN'S ILLNESS

After I left Cambridge, in the beginning of 1918, I did not have very many opportunities of meeting Ramanujan; but when I heard from one of our common friends that he was ailing and in hospital, I called on him at the hospital. Ramanujan was being nursed in bed for T B but I doubt if he had been informed that his was a hopeless case. He did not appear unduly down-cast and told me, I think, that he was looking forward to returning to India as soon as he was in a fit condition to travel. But his condition steadily worsened and the fell disease was very soon to carry him away and deprive the country of its most outstanding mathematical genius.

MN Reminiscences of Dr P C Mahalanobis F R S, Member of the Planning Commission of India.

MN1 RAMANUJAN'S INTUITION

I joined King's College, Cambridge, in October 1913. I was attending some mathematical courses at that time including one by Professor Hardy. A little later, we heard that S Ramanujan, the mathematical prodigy, would come to Cambridge. I used to do my tutorial work with Mr Arthur Berry, Tutor in Mathematics of King's College. One day I was waiting in his room for my tutorial, when he came in after having taken a class in elliptic integrals.

BERRY: Have you met your wonderful countryman, Ramanujan?

MAHALANOBIS: I have heard that he has arrived; but I have not met him so far.

BERRY: He came to my elliptic integrals class this morning. (This was sometime after full term had begun, and I knew Mr Berry had already given a few lectures on that subject.)

MAHALANOBIS: What happened? Did he follow your lectures?

BERRY: I was working out some formulae on the blackboard. I was looking at Ramanujan from time to time to see whether he was following what I was doing. At one stage, Ramanujan's face was beaming and he appeared to be greatly excited.. I asked him whether he would like to say anything. He then got up from his seat, went to the blackboard, and wrote some of the results which I had not yet proved. Ramanujan must have reached those results by pure intuition. His facility in the theory of numbers was in a large measure intuitive. He made numerous conjectures, like other pure mathematicians. Many of the results apparently came to his mind without any effort. He was, however, aware that a good deal of intellectual effort would be required to establish his philosophical theories.

80

MN2   GETTING INTO BED

I was fortunate in forming a good friendship with Ramanujan very soon. It came about in a somewhat strange way. One day, soon after his arrival, I went to see Ramanujan in his room in Trinity College. It had turned quite cold. Ramanujan was sitting very near the fire. I asked him whether he was quite warm at night. He said that he was feeling the cold though he was sleeping with his overcoat on and was also wrapping himself up in a shawl. I went to his bedroom to see whether he had enough blankets. I found that his bed had a number of blankets but all tucked in tightly, with a bed cover spread over them. He did not know that he should turn back the blankets and get into the bed. The bed cover was loose; he was sleeping under that linen cover with his overcoat and shawl. I showed him how to get under the blankets. He was extremely touched. I believe this was the reason why he was so kind to me.

MN3   APPLIED MATHEMATICS

On Sunday mornings, Ramanujan and I often went out for long walks. One Sunday it had been arranged that we would both have our breakfast in my room and then go out for a walk. It was a cold morning with some snowfall. I was a bit late in getting up and was shaving in my bedroom when he arrived. I asked him to wait in the sitting room. When I came out, I found that he was reading Loney's *Dynamics of a particle* with great interest. Seeing me, he put back the book on the table and said that it was very interesting. Evidently, he had never studied dynamics but had got interested in what he was reading.

MN4   RAMANUJAN'S FLASH

On another occasion, I went to his room to have lunch with him. The First World War had started sometime earlier. I had in my hand a copy of the monthly *Strand Magazine* which at that time used to publish a number of puzzles to be solved by readers. Ramanujan was stirring something in a pan over the fire for our lunch. I was sitting near the table, turning over the pages of the *Magazine*. I got interested in a problem involving a relation between two numbers. I have forgotten the details; but I remember the type of the problem. Two British officers had been billeted in Paris in two different houses in a long street; the door numbers of these houses were related in a special way; the problem was to find out the two numbers. It was not at all difficult. I got the solution in a few minutes by trial and error.

MAHALANOBIS: (In a joking way), Now here is a problem for you.

RAMANUJAN: What problem, tell me. (He went on stirring the pan).

I read out the question from the *Strand Magazine*.

RAMANUJAN: Please take down the solution. (He dictated a continued fraction).

The first term was the solution which I had obtained. Each successive term represented successive solutions for the same type of relation between two numbers, as the number of houses in the street would increase indefinitely. I was amazed.

MAHALANOBIS: Did you get the solution in a flash?

RAMANUJAN: Immediately I heard the problem, it was clear that the solution was obviously a continued fraction; I then thought, "Which continued fraction?" and the answer came to my mind. It was just as simple as this.

MN5 PRODUCT OF ZERO AND INFINITY AND PHENOMENAL WORLD

I have mentioned that Ramanujan and I often used to go out for long walks on Sunday mornings. During these walks our discussions ranged over a wide variety of subjects. He had some progressive ideas about life and society but no reformist views. Left to himself, he would often speak of certain philosophical questions. He was eager to work out a theory of reality which would be based on the fundamental concepts of "zero", "infinity" and the set of finite numbers. I used to follow in a general way, but I never clearly understood what he had in mind. He sometimes spoke of "zero" as the symbol of the Absolute (*Nirguna-Brahmam*) of the extreme monistic school of Hindu philosophy, that is, the reality to which no qualities can be attributed, which cannot be defined or described by words, and which is completely beyond the reach of the human mind. According to Ramanujan, the appropriate symbol was the number "zero", which is the absolute negation of all attributes. He looked on the number "infinity" as the totality of all possibilities, which was capable of becoming

82

manifest in reality and which was inexhaustible. According to Ramanujan, the product of infinity and zero would supply the whole set of finite numbers. Each act of creation, as far as I could understand, could be symbolised as a particular product of infinity and zero, and from each such product would emerge a particular individual of which the appropriate symbol was a particular finite number. I have put down what I remember of his views. I do not know the exact implication. He seemed to have been perhaps emotionally more interested in his philosophical ideas than in his mathematical work. He spoke with such enthusiasm about the philosophical questions that sometimes I felt he would have been better pleased to have succeeded in establishing his philosophical theories than in supplying rigorous proofs of his mathematical conjectures.

MN6 RAMANUJAN'S PERSONALITY

Ramanujan had a somewhat shy and quiet disposition, a dignified bearing, and pleasant manners. He would listen carefully to what other people were saying but would usually remain silent. If he was asked any question, or on rare occasions, if he joined in any general conversation, he would speak frankly, but briefly. Whilst speaking to a friend or in very small groups, he would, however, expound his own ideas with great enthusiasm, not only on philosophical questions but occasionally also on other subjects in which he was seriously interested. Although I could not follow his mathematics, he left a lasting impression on my mind. His bright eyes and gentle face with a friendly smile are still vivid in my mind.

MP/MS NON-CLASSMATE FRIENDS

MP Reminiscences of Mr K Gopalachary, retired from the Revenue Board Office, Madras

MP1 GOD, ZERO, AND INFINITY

As one talked to Ramanujan, one would forget oneself. In 1909, I was living with my brother K S Patrachariar, Lecturer in Mathematics in the Teachers' College, Saidapet. It was 8 one evening. Some of the post-graduate students of my brother had come to our house. Ramanujan also arrived. He began a talk on "God, Zero, and Infinity". All of us were spell-bound. The company broke off only at 2 a m. Next morning, I had a talk with my brother on this.

GOPALACHARY: I felt dazed last night. I could not follow his talk much. What did he say?

PATRACHARIAR: While listening to him, I thought I was following him. But now it is all like a dream. I am not able to recall anything coherently.

MP2 HOW IT ALL BEGAN

GOPALACHARY: Ramanja, how did you get so much knowledge in Mathematics?

RAMANUJAN: It got started in different ways. I shall tell you one of them. One night, I had the following dream. I heard a pedlar's voice in the street. He was selling pills. The price of each pill was less than one anna (present six Paise or two American Cents). But the price of only one was 50 Paise. I asked him what its special use was. He said that he did not know about it. I bought it immediately. The next day the ideas about Arithmetic Progression, Geometric Progression, and Harmonic Progression began to develop in my mind.

MP3 RANGE OF INTEREST

In spite of such uncanny experiences on one side, he used to enjoy subtle intellectual discussions—particularly on the Indian Schools of Philosophy. One day we both went to attend a discourse on the atheism in Sankhya Philosophy, by Marur Ganesa Sastri.

THE LECTURER: If there is a beginning, there should be an end, and if there is no beginning, there will be no end. *Karma* (Casual-Action—Chain) has no beginning.

RAMANUJAN: (Jumping up spontaneously) If that be so, is there no *Moksha* (Release from the cycle of births and from Casual-Action—Chain) at all, according to this School of Philosophy?

The lecturer could not answer this question. But Ramanujan explained how the *Moksha* idea and the *Karma* idea could be reconciled. The audience were struck by his ability.

84

MP4 OCCULT EXPLANATION

Ramanujan used to give occult explanations of Planets, Stars, Colour Blindness, and so on. He was deeply interested in occult subjects such as Life After Death, Astrology, and Psychic Phenomena. Here are two examples.

MP41 Vision of Highly Evolved Souls

Satyapriyarayar was his old drill master at school. Later on he went mad. As he became unruly, he was chained. By the time he recovered, he had lost his appointment. Even daily food became a problem for him. Ramanujan used to collect food from house to house and give it to him.

GOPALACHARY: Why do you worry yourself about this mad man?

RAMANUJAN: This drill master was one day sitting in the inside verandah of his house. He suddenly experienced an illusion of a number of Lilliputian-like creatures standing all round him. This terrified him and threw his mind out of order. In reality, on account of the merits earned by him in his earlier births, he really had a vision of highly evolved beings. That is why I respect him.

MP42 Space-Time-Junction-Point

When he was studying in Pachaiyappa College, Ramanujan was a tenant in a house in George Town. One day he dreamt as follows: A child of a family living in another room in the same house appeared to be in danger of death. However, when he got up in the morning, the child was quite normal. But, when Ramanujan returned from College, he heard that the child had high fever. Within two days, the temperature rose very high and the child became unconscious. On seeing this, Ramanujan told his parents, "Please remove the child to the next house. For, the death of a person can occur only in a certain space-time-junction-point. You should avoid this house for some days." The parents removed the child to the next house. A few minutes later, the child recovered consciousness and insisted on his being taken back to his room. But Ramanujan pleaded with the parents not to do so. However, the parents did not give heed to his words. They took the child back to their house. And after a short while, the child expired. Later on, in his dream Ramanujan met that child. Immediately Ramanujan was seized with fear. He therefore

began to run away from that place.   But the child pursued him.
On reaching a temple, Ramanujan entered it, thinking that the
child could not enter it.   But sometime later, other devotees
coming into the temple raised a hue and cry saying that there was
the dead body of a child outside the temple.   This made Ramanujan
feel that he need not fear any longer.   But when he came out,
the child sprung on him.   At this stage Ramanujan woke up.
He found that he was shaking with fear.   The next day he developed very high temperature.   Medicine would not bring it down.
Thereupon, Ramanujan prayed sincerely to Goddess Namagiri.
Shortly afterwards he fell into a sleep. Goddess Namagiri appeared
in his dream and told him, "Drink the water in which boiled rice
has been kept over-night."   On waking up, he carried out the
direction of Goddess Namagiri.   He then recovered.

MP5   EQUABLE TEMPER

I had never seen him lose his temper however badly anybody
contradicted or even ridiculed him.   He was quite indifferent to
all such things.   I do not remember of his speaking of sex, at
any time.

MP6   THE FAMOUS NOTEBOOK

When Ramanujan first took his famous Notebook to a Professor
of Mathematics in 1910, it appears that he turned through the
pages and then dashed it on to Ramanujan.   He came back to
our house quite unmindful of that insulting behaviour and left
that Notebook in our house.   It was lying in our store-room for
a long time.   R Swaminatha Ayyar, an old student of G H Stuart,
a former Professor of Mathematics of the Presidency College,
Madras, happened to go through it.   He was astonished at its
contents, and mentioned about it to B Hanumantha Rao.

MP7   MATHEMATICS UNDER A COT

One afternoon a cousin of mine and myself went to his house
to take him out for a walk.   There I found him sitting in the dark
under a cot and writing on his slate.   My cousin explained to
me the reason for his sitting under the cot, as follows: "He is
always engaged in doing something of mathematics.   His parents
do not like his doing so, without doing anything by way of earning.

86

Therefore, he hides himself under the cot whenever he has to do his mathematics."

MP8 SANKHYA

I met him only twice after he returned from England. On one occasion, he said that he was keen to read the Atheistic Text of Sankhya once more. (*See* Sec MP3).

MQ Reminiscences of Mr T K Rajagopalan, Retired Accountant General, Madras.

MQ1 ELLIPTIC INTEGRALS IN DREAM

I had known Ramanujan when he was a boy in Kumbakonam. Before 1911, I was Assistant Accountant General in Madras for a few years. I was living in a bungalow in Purushawakkam. Ramanujan was then unemployed. He came to me one evening and wanted to sleep in my house. But all the rooms in the main building were occupied. There was only one outhouse. A monk was occupying it. I asked Ramanujan whether he would mind sleeping in the outhouse along with the monk. He welcomed it. Next morning he came and told me, "He is such a powerful soul, I am glad I had the opportunity to share the room with him. His presence stimulated me a good deal. While asleep I had an unusual experience. There was a red screen formed by flowing blood as it were. I was observing it. Suddenly a hand began to write on the screen. I became all attention. That hand wrote a number of results in elliptic integrals. They stuck to my mind. As soon as I woke up, I committed them to writing."

MQ2 NARASIMHA

I am giving here an extract from my book *Hidden Treasures of Yoga*, which relate to Ramanujan. "Ramanujan and his family were ardent devotees of God Narasimha (the lion-faced incarnation (*avasara*) of God), the sign of whose grace consisted in drops of blood seen during dreams. Ramanujan stated that after seeing such drops, scrolls containing the most complicated mathematics used to unfold before him and that after waking, he could set down on paper only a fraction of what was shown to him.

MR **Reminiscences of Mr R Srinivasan,** Retired Professor of
Mathematics, Trivandrum.

MR0 UNKNOWN DETAILS

I am giving here some information and incidents about
Ramanujan, either not known or not recognised.

MR1 RELIGIOUS PRACTICES

Very few people had opportunities to know him as a man, to
know his inner make-up and his attitude to life.   He was a mystic,
a true mystic in the full significance of the term.   He was intensely
religious, almost superstitious in some daily observances — he
would not take food unless cooked by people he approved; he
was a very strict vegetarian.   While in England he cooked his
own meal.   Some persons said that this was largely responsible
for the fatal failure in his health.   I am not quite so
sure.

MR2 MYSTIC

That he was highly intuitive and got at the truth of things, in
a flash, cannot be denied.   He saw the truth and knew it, though
he found it difficult to explain to others in terms of logical sequence.
When I was in Trivandrum in 1913, I used to go to Madras often.
Ramanujan, then a clerk in the Madras Port Trust, somehow
took a fancy to me and used to visit me whenever I was in Madras;
perhaps he found a sympathetic listener to what he wished to
say. He used to show his notes to me, but I was rarely able to make
head or tail of at least some of the things he had written.   One
day he was explaining a relation to me; then he suddenly turned
round and said, "Sir, an equation has no meaning for me unless
it expresses a thought of GOD."   I was simply stunned.   Since
then I had meditated over this remark times without number.
To me, that single remark was the essence of Truth about God,
Man, and the Universe.   In that statement I saw the real
Ramanujan, the philosopher-mystic-mathematician.

MR3 PRE-COGNITION

He had also other strange experiences — what people would call
supernatural.   He could foresee events as in a vision. Finding

88

a sympathetic spirit in me he used to mention even some of his intimate experiences. Let me mention just one of them. One day he saw a vision. He was in a house he had not seen before. Under a pillar in one of the verandahs he saw a distant relative of his lying dead with people in mourning all round. The vision then vanished and he forgot all about it. Sometime later he happened to visit a relative who was then employed in some town far away. Imagine his surprise when he found that the house of his relative was exactly the house he saw in his vision years ago. When he was talking to his relative, he learned that there was a patient who was undergoing medical treatment and was staying in the house. It was a few days later that he saw that patient lying in a bed under the very pillar of his vision; it was the same person he saw in the vision. He was taken aback; in a few moments the patient died under that very pillar.

MS Reminiscences of Mr S Venkatarama Ayyar, Advocate, Madras.

MS1 MAYA
Ramanujan would often be in a contemplative mood. He and I were returning to Triplicane in the evening in a tramcar, sitting on the long bench at the driver's back. The driver was alternately accelerating the speed or applying the brakes with gusto. Ramanujan burst out, "See how that man is imagining that he has the power to go slow or fast at his pleasure. He forgets that he gets the power through the current that flows in the overhead wires, which is not visible to him unless he tries to *see* it. That is how *maya* works in this world."

MT Reminiscences of Mrs Janaki Ramanujan (with acknowledgements to *The Hindu*).

MT1 MY FIRST YEAR WITH MY HUSBAND'S FAMILY
In our days girls were married even when they were children. Indeed I was married when I was only nine. Obviously I was too young to join my husband and begin my wedded life. However, in conformity to the then prevailing custom, I used to go and live in his home for some months. In 1913, my husband and his

grandmother were living in Saiva Muthiah Mudali Street, George Town. They had selected that house because it was near the Port Trust Office where my husband was working. I joined the family at that time. Later we moved into a house in Hanumantharayan Koil Street in Triplicane. We again changed to Thope Venkatachala Chetty Street in Triplicane itself. Years later it became the home for a Music school. My husband left for Cambridge when we were living in that house.

MT2 TO MAKE BOTH ENDS MEET
The members of the family were too many for my husband's monthly salary of Rs 25. Therefore he was giving tuition in Mathematics to some college students to make both ends meet. He was thus kept busy most of the day. He was unassuming and simple in his habits. My experience in the family at that time, though brief (less than a year), was memorable and could not be forgotten.

MT3 ABSORPTION IN THOUGHT
He was often deeply engrossed in his Mathematics and would even forget his food. This became quite common after he became a research scholar in the University. He had to be reminded about his food. On some occasions his grandmother or mother would serve, in his hand, food made of cooked rice mixed with sambhar, rasam, and curd successively. This they did, so that his current of thought might not be broken. He would ask his mother or grandmother to wake him up after midnight so that he could go on with his work in the silent and cooler hours of the after-night.

MT4 LIFE IN ENGLAND
When he was in Cambridge, he used to write to his parents saying that he continued to observe the prescribed religious practices without any break. He prepared his own food. After his return home, he told me that he felt very happy when the famous Editor of *The Hindu*, Mr Kasturiranga Ayyangar came to his room and partook with relish the *pongal* (a South Indian delicacy prepared with rice, greengram, ghee, and a dash of spices) prepared and served by him.

90

MT5 RAMANUJAN AS ASTROLOGER

My husband was good in Astrology. Many of his relatives and friends used to come often to consult him about their future and particularly for fixing auspicious hours for different festive or religious functions. I used to see this happening before he went to England.

MT6 HIS LAST DAYS

He returned from England only to die, as the saying goes. He lived for less than a year. Throughout this period, I lived with him without break. He was only skin and bones. He often complained of severe pain. In spite of it he was always busy doing his Mathematics. That, evidently helped him to forget the pain. I used to gather the sheets of paper which he filled up. I would also give the slate whenever he asked for it. He was uniformly kind to me. In his conversation he was full of wit and humour. Even while mortally ill, he used to crack jokes. (*See* Sec N22). One day he confided in me that he might not live beyond thirty-five and asked me to meet the event with courage and· fortitude. He was well looked after by his friends. He often used to repeat his gratitude to all those who had helped him in his life.

MT7 MEMENTOS

We did not have children. I have only two mementos with me. One consists of the two vessels in which I used to warm water for fomentation whenever he complained of acute pain. The second consists of two portraits of himself.

RAMANUJAN, THE MAN

N1 The Eyes

The eyes of a person are one of the windows through which the mind receives information from outside the skin. In this they have the same status as all the other sensory organs. But they also have quite another function. This is something distinctive. They are also the windows through which we can get a glimpse of the mental and spiritual qualities of the person. These qualities are usually hidden away by the physical sheath of the person — whether it be handsome or ugly or anything between these two extremes. The eyes of Ramana Maharishi unmistakably disclosed his self-realisation. The poet Valmiki highlights this important function of the eyes in several places. For example, Hanuman — a stranger to Sita — goes to her with a message from Rama when she is in captivity in far off Lanka. To make sure of the genuineness of the messenger, Sita asks Hanuman to describe the physical features of Rama. In doing so, she puts questions about his arms, his body, his legs, and so on. Hanuman takes it as a clever way of questioning in order to trap him. Instead of describing these, he begins with the description of Rama's eyes and says, "When one sees Rama, who will think of any of his organs except his eyes?" The eyes truly throw into the shadow, as it were, all the other physical features of a person whether he is handsome or ugly. In the case of Rama, an exquisitely handsome body was suppressed from one's attention by the beauty of his eyes. Also in the case of Ramanujan. All his physical features escaped our attention, whenever we were face to face with him, conversing with him or listening to a talk by him. It were the eyes alone that would engage our mind. The glow in them would captivate us. The glow would increase in its charm and intensity, as he warmed up in his conversation or in his talk. Indeed the eyes showed Ramanujan, the Man.

92

N2 Personal Qualities

N21 SHYNESS

Ramanujan was of a shy temperament. His face showed that
he was contemplative. This enhanced the impression about his
shyness. But, when in the midst of intimate friends, he was
very participative. He would enjoy discussion with them. If
the topic was philosophical or one concerning Indian classics,
he would warm up to a remarkable degree. On such subjects,
he would spontaneously express his ideas without any trace of
shyness even in the midst of strangers. Again, while expounding
the mystic significance of numbers, he showed no shyness. In
spite of all this, Ramanujan generally passed for a shy person.

N22 WIT AND HUMOUR

The reminiscences of his college friends, given in Chap M,
show that he was a man with a great sense of humour. His wit
also was remarkable. His wit and humour did not desert him
even when he was laid up with tuberculosis in the last year of his
life. Mrs Ramanujan disclosed to me a few grim instances of
this. After his return home in 1919, he lived only for about a
year. He was extremely ill with tuberculosis. He was confined
to bed. He spent the first few months in Kodumudi, a village
on the banks of the river Kauveri. The doctors felt that the place
was too humid for a tuberculosis patient. They recommended his
living in Tanjavur, a relatively dry town. The Government and
the University of Madras insisted on his going to Tanjavur. On
hearing this, he punned on the word 'Tanjavur' and broke it into
the three elements 'tan', 'savu' and 'ur' and said to his wife, "They
want me to go to tan-savu-ur — the place of my death!" This
is the literal meaning of the Tamil words in that broken form.
Later, the Government and the University had him shifted from
his hometown Kumbakonam to Chetpet, a Division of the City
of Madras. On reaching that place, he punned on the name of
the Division and broke it into the compound Tamil word, 'Chat-
pat' and told his wife, "They have brought me to the place where
everything will be 'chat'-'pat'. In this broken form, the Tamil
word means, "It will happen very soon." What a grim wit to
come from a dying man! Such was Ramanujan, the Man.

N23   Unperturbed Emotion

It was impossible to offend Ramanujan. He could never be driven to anger or bitterness or complaint. No amount of teasing or offensive conduct on the part of others would disturb his equanimity. Some people — usually those who were incapable of controlling their own passion — used to describe this equanimity of Ramanujan as insensitiveness and callousness. Even this did not offend him. He used to say, "They speak in their ignorance." That was Ramanujan, the Man.

N24   No Disgust with Life

Ramanujan had to spend the first twenty-three years of his life in poverty. His parents could not understand him as he was absorbed most of the time in working out something on his slate and copying out the result in a notebook, quite unmindful of the need to get employed and earn. They were annoyed with him. None of these factors could take him away from enjoying his own life, however exclusive it was. He was never disgusted with life. Even in the worst condition — as it was when he had to run away from home in 1905 — he did not do it in disgust of life. He did so only to escape temporarily from the environment in which he was not correctly understood.

N25   Emotional Warmth and Delight

The emotion of Ramanujan would warm up when he was expounding strange metaphysical theories and explaining the phenomenal world on the basis of such theories. Some felt that his emotions did not get that warmth when he did his mathematics. Probably, it was more than warmth of emotion. It was delight that enveloped him while reaching certain depths in mathematics.

N26   Friendliness

Even from his childhood, Ramanujan had a small circle of friends. They enjoyed his unusual mathematical powers and philosophical acumen. Many of his contemporaries at school and college have testified to this. His seniors in mathematics have also found in him a likeable friend. They enjoyed his company whether it was in the pursuit of mathematics or in conversation about philosophical and religious thought and experiences. Even at a lower level, one could enjoy his company over the tea table and discuss Mathematics, Mysticism, Philosophy, Religion, or Politics.

N27 MODESTY

The modesty and unassuming quality of Ramanujan struck everybody who moved with him after he attained fame as a great mathematician. In this, he seems to have been an embodiment of *Hri*—the quality of modesty. According to Hardy, when the Trinity College Fellowship of £250 a year was announced to him, Ramanujan felt embarrassed and worried. In effect he told Hardy, "How do I deserve it? In this health how can I justify accepting it?" Hardy had to say, "You have done enough already. What you have done deserves much more than this. This is given to you to enable you to live in comfort. You may reside wherever you like. There is no obligation of any kind either by way of teaching or by way of doing any further research." In fact, he had to be reminded of the greatness of his own achievement even as Sri Rama had to be after winning the war in Lanka, first by the Creator Brahma and a few days later by the sage Bhardwaja, when Sri Rama went to pay his respects to him at Allahabad on his way back to Ayodhya.

N3 Industry

The industry of Ramanujan was prodigious. To read through the dry-as-dust collection of formulae in Carr's *Synopsis* required not only intellectual capacity but also abundant industry. Ramanujan had it even when he was at school. After he left college, he was incessantly working—not to earn money, but to find out the truth and the beauty of mathematics. He worked most of the day and night. He worked even when Death was knocking at his door.

N31 FAVOURITE OF MAHASARASWATI

According to Hindu tradition, he was fully endowed by Mahasaraswati. His industry was the demonstration of the power of Mahasaraswati as described in Sri Aurobindo's *Mother*. "Mahasaraswati presides over the details of organization and execution, relation of parts and effective combination of forces and unfailing exactitude of result and fulfilment.

The science and craft and technique of things are Mahasaraswati's province. Always she holds in her nature and can give to those whom she has chosen, the intimate and precise knowledge the subtlety and patience, the accuracy of an intuitive mind and conscious hand and discerning eye of the perfect worker. This power is the strong, the tireless, the careful and efficient builder, organizer, administrator, technician, artisan, and classifier of the worlds. When she takes up the transformation and new building of the nature, her action is laborious and minute and often seems to our impatience slow and interminable, but it is persistent, integral and flawless. For the will in her works is scrupulous, unsleeping, indefatigable; leaning over us she notes and touches every little detail, finds out every minute defect, gap, twist or incompleteness, considers and weighs accurately all that has been done and all that remains still to be done hereafter. Nothing is too small or apparently trivial for her attention; nothing however impalpable or disguised or latent can escape her. Moulding and remoulding, she labours each part till it has attained its true form, is put in its exact place in the whole and fulfils its precise purpose. In her constant and diligent arrangement and rearrangement of things, her eye is on all needs at once and the way to meet them and her intuition knows what is to be chosen and what rejected and successfully determines the right instrument at the right time, the right condition and the right process. Carelessness and negligence and indolence, she abhors; all scamped and hasty and shuffling work, all clumsiness and a *puepres* and misfire, all false adaptation and misuse of instruments and faculties and leaving of things undone or half done is offensive and foreign to her temper. When her work is finished, nothing has been forgotten, no part has been misplaced or omitted or left in a faulty condition; all is solid, accurate, complete, admirable. Nothing short of a perfect perfection satisfies her and she is ready to face an eternity of toil if that is needed for the fulness of her creation. Therefore of all the Mother's Powers she is the most long-suffering with man and his thousand imperfections. Kind, smiling, close and helpful, not easily turned away or discouraged, insistent even after repeated failure, her hand sustains our every step on condition that we are single in our will and straightforward and sincere; for, a double mind she will not tolerate and her revealing irony is merciless to drama and histrionics and self-deceit and pretence. A mother to our wants, a friend in our difficulties, a persistent and tranquil counsellor and mentor, chasing away with her radiant smile the clouds of gloom and fretfulness and depression, reminding always of the ever-present help, pointing to the eternal sunshine, she is firm, quiet and persevering in the deep and continuous urge that drives us towards the integrity of the higher nature." Ramanujan's industry and his success through industry makes us feel that he was a favourite of Mahasaraswati.

N4   Intellect and Intuition

Apprehension of knowledge and its furtherance can be achieved with the help of the primary senses and the intellect. These two are generally inseparable. We shall use the term 'Intellect' to denote the two taken together. Apprehension of knowledge can also be achieved without the mediation of the intellect. The mediating agency in this case, we shall denote by the term 'Intuition'. According to Indian tradition, Knowledge through intuition is thing-dependent and not doer-dependent. To use the Indian terms, it is *Vastu-tantra* and not *Kartru-tantra*. Intellection is *Kartru-tantra*, that is, doer-dependent. Intellection can never lead to Total Apprehension. It will always be partial. Even in the case of the best intellectuals, it can at best make only a very close approximation to Total Apprehension, but cannot reach it. In the case of a full-blown intuitionist, intuition will give Total Apprehension. Intuition is Trans-Intellectual. And according to Indian usage, an intuitionist 'sees' and does not approximate through the intellectual processes of induction, deduction, and proof.

N41   INTUITION AS LIMITING POINT OF INTELLECT

Most thinkers are intellectuals. There may be a dash of intuition in the case of the abler among them. Their intuition will not be pure at all times. It may be pure only for a split second, perhaps. The knowledge obtained through intuition during that short while will be totally trustworthy. But when intuition fades and is weak and the intellect steps in, it will not be so. In that case, intellect will have to test and verify such a piece of knowledge in its own

97

laborious way.   Most of the well-known leaders of knowledge
in diverse subject fields belong to this category.   It is difficult to
say which of their work was the result of pure intuition and which
was not.   It is only one with full intuition that can determine it.
Intellectuals make errors in assessing the nature of their results,
whether it originated from intellect or from pure intuition itself.
For, the progression of the intellect approaches the intuition as
the limiting point.   But the range of the intellect is an 'open-set'—
that is, it does not include the limiting point.

N411   *Flair and Intuition*

One should not mistake a high speed of intellect for intuition,
because intuition is instantaneous.   It cannot be acquired.   One
is born with it.   On the other hand, speed in intellectual work
can be increased to any extent by incessant practice and undivided
attention and concentration.   Nor should one mistake intellectual
flair for intuition.   Intuition is non-dependent on prior sensory
and intellectual experience.   On the other hand, flair sprouts
from intellect soaked, as it were, in all related experiences at the
conscious and subconscious levels.   In the intellectual process,
there is direction towards a vaguely seen or preconceived result.
In intuition, there is no such direction.   The result is simply 'seen'
without effort and without expectation.   Perhaps, this is an over
simplification of the problem made by intellect.   Intuition alone
can distinguish between flair of a very high order and pure intuition.
Perhaps the distinction is ineffable, in the language of intellection.

N42   MIXTURE OF TRUE AND FALSE CONJECTURES

In the case of Ramanujan, the incidence of pure intuition should
have been more frequent than in the case of most mathematicians.
Probably its duration also was longer.   But neither he nor others
can say when pure intuition was working and how long.   That
is the reason, perhaps, for some of his conjectures being true
and some false.

N43   SUDDEN CHANGE-OVER

In the case of a person rich in intuition, there may be a sudden
change-over from intellection to intuition, even in the middle of
a particular pursuit.   This happened in the case of Sankara.   He

was standing in front of the deity Subramanya in Tiruchender, a coastal town in Tinnevelly District, Madras. There is a sudden change-over of the style and the flow of the hymn. Sankara himself felt it. His description of it occurs in the middle of that very hymn. "No more stilted partial description of the Lord, catching word after word by laborious search through the darkness of the intellect. The light of intuition has now begun to bathe the personality of the Lord. Hereafter, I must merely describe what I *see*."

N44   TASK OF PROVING

After going to Cambridge, Ramanujan was put on the way of "proving" also – and not merely "verifying" — both pieces of intellectual work. It has been said that it is an incorrect use of a person rich in intuition to make him take off his coat, as it were, and to supply proof for everyone of his conjectures, so as to eliminate the chance for false conjectures. It would be like making it obligatory for a dowser himself to dig a well on the spot and make sure that the result of this dowsing was correct. A measure of the trustworthiness of one conjecturing in mathematics and of a dowser is the probability of its standing the test of proof. This probability was unusually high in the case of Ramanujan. I think it was Littlewood that said a proper use of Ramanujan would be to leave him to his intuition and to his daring, penetrating conjectures, since such persons are few and far between. The proof and the verification is the work for a technician in mathematics. It is necessary work, no doubt, and persons should be trained for this technician's work. The purpose of departments of mathematics is to train such technicians. It is of the very essence of training in mathematics. It would have been bad economy, as Hardy said, to have tamed and trained Ramanujan in that way and made a technician of him. Intuition characterised Ramanujan not only as mathematician but also a Man. This was felt by his friends when he was giving religious, philosophical, or any other non-mathematical expositions to small intimate groups.

N5   Religious Experience and Philosophy

Ramanujan's religious experiences and his philosophy were not easily reconcilable. It is so, in the case of anybody with religious experience, as distinct from conformity to religious rituals. Religious experience belongs to the domain of intuition. It is concerned with the experience of God. Philosophy, on the other hand, belongs to the domain of intellect. It seeks to establish or deny God by intellection. In philosophy he was fascinated by the Atheistic Sankhya School of Philosophy. On the other hand, his behaviour was as if he had experienced God — not the God, the Absolute — but a manifestation of God. The manifestation nearest to him was Namagiri. In the phenomenal life one has often to lean upon some such manifestation. If the manifestation is embodied as a Man, it is called *Avatara* in Indian tradition. If it is not an *Avatara*, it is called an *Avasara*. Both Rama and Krishna as *Avataras*, and Namagiri and Narasimha as *Avasaras* of the Absolute, were meaningful to Ramanujan. To others other manifestations will be meaningful. For Sadhu Sunder Singh the favourite manifestation was as Jesus Christ. For Ramakrishna Paramahamsa it was Kali. Ramana Maharishi on the other hand was in communion with the Absolute Itself. Each denominational religion has a particular manifestation as its basis. Whatever be the manifestation, the religious experience is the realization of the Absolute. It is in this sense that Ramanujan should have told Hardy that all religions seemed to him more or less equally true. Unfortunately, this should have led Hardy to infer that Ramanujan's religion was a matter of observations and not of intellectual conviction. Hardy even suspected that he was an Agnostic. In fact, religious experience is not a matter of phenomenal observations or of intellectual experience. It is trans-phenomenal and trans-intellectual. However, in phenomenal life one always leans upon some manifestation of God and some rituals. Even in the highly sophisticated intellectual world of Universities, we do lean upon a ritual like convocation and of an observance like wearing hood and gown. Perhaps the best testimony to the dual life of man in the world of religious rituals and that of religious experience is well brought out in a hymn of Sankara. In this hymn, Sankara addresses the Absolute — the one God as one would call it — in the words, "My God, I have called you the nameless by several names. I have offered you flowers which are only your own manifestations. I have offered you food which is only your own manifestation. To facilitate my perambulation round you, I have built a temple to imprison you who is omni-present. Above all, hymns are being uttered in praise of you, by me who is only a manifestation of you." Religious observance belongs to the phenomenal world. It cannot be easily escaped by any, except by a soul of near-realization such as Ramana Maharishi or Ramakrishna Paramahamsa. Even they provided for religious observances out of consideration for the common man, whose religious experience is nil or slight, but who can find solace only in religious observances. Ramanujan's dualism — equating all religions on the one hand and keeping to a specific set of religious observances — is a normal characteristic of Ramanujan, the Man.

N51   METAPHYSICS IN SYMBOLS

Ramanujan's symbolic Metaphysics is of interest. He would symbolize God and the existents or entities of the phenomenal world symbolically as follows:

God as the Absolute is Attributeless (Nirguna-Brahman). In this view, 'Zero' may be taken to represent God.

God is also the Abode of all Attributes (Sarva-Guna-Asraya). As such God is Sa-Guna. In this view 'Infinity' may be taken to represent the infinity of attributes found in and along with their Abode.

The combinations of the Infinity of Attributes, taken any one or any two or any three, etc. at a time, is itself infinite.

Each combination of the Absolute and of one of the Infinity of combinations of the Attributes appears to us as an existent or entity in the phenomenal world.

This corresponds in symbols to the product of 'Zero' and 'Infinity' being indeterminate and thus admitting itself to be taken to yield any one of the infinity of numbers.

Thus, the phenomenal world of the past, present, and future is represented by the product of 'Zero' and 'Infinity'.

This symbolic representation by Ramanujan of God and of the phenomenal world, has been indicated in the respective reminis-cences of Gopalachary in Sec MP1 and of Mahalanobis in Sec MN5.

N6   Occultism

Everybody does not have occult or psychic experience. Many

are insensitive to it. A few, however, are sensitive to it. Ramanujan belonged to the latter class. Many instances of his occult experiences are given in other sections. The reminiscences of K Gopalachary given in Sec MP4 are particularly rich in such experiences. It is out of bounds for a person insensitive to occult phenomenon to decry the occult experiences of one sensitive to it. At best he can only say, "I don't understand them." Leaning towards Occultism was a characteristic of Ramanujan, the Man.

N7 Social Practices

Ramanujan was a conformist in social practices — be they group practices or personal ones. For example, he was a strict vegetarian. He would change his pyjamas before starting to cook. He wanted to assure his parents about his conformity in this and other daily practices. Therefore in his letters to his parents, he used to mention his conformity to the practice of his social group. But due to the impact of Western culture on Indian culture during the last two centuries, he was made to adopt alien practices in his dress, treatment of hair, and certain other matters. This caused him pain and dislike in his parents. Some pity this; some others ridicule it; while still others praise it. None of these evaluations can be justified. For, social practices are inherited by a person. Under social pressure and the pressure of imitation they get woven into the habits of a person before reaching adulthood. Thereafter when exposed to the clash of divergent practices as a result of the social impact of different cultural groups, the reaction may vary from person to person. It is severely personal. It is immaterial whether he conforms to his inherited social practices or changes over to alien ones. Some may adopt the principle, "When you are in Rome do as the Romans do." The guiding principles for others may be, "Follow your own habit, wherever you happen to go." Comfort also may indicate different decisions to different persons. Some will decide on the basis, "Let not the difference in my habits attract the attention of the new social group; then only I can do my work with comfort and efficiency." Others may decide on the basis, "I can have the greatest comfort and do my best only if I continue my established habits. If I change over to the current practices of the social group into which I am thrown temporarily, my mind will always feel distracted and to that extent my efficiency will be affected." Either basis for decision may be at variance with what reasoning might indicate. Moreover brisk international travel and cultural mix-up are now rapidly changing the social habits of most communities. This again shows that there is neither vice nor virtue in either being a no-changer or a willing-changer in our personal habits. Ramanujan happened to be a no-changer. It must be stated here that social practices and religious practices get blended and become indistinguishable.

N8 Gospel of Wealth

Ramanujan's attitude towards wealth was unusual. He had been without money to satisfy the normal wants till he was 26 years of age. Even then, he found it undignified to accept a gift of money from others. From his twenty-sixth year, he had income sufficient to live in comfort. In 1918, he had an assured income of £500 a year — £250 from the Trinity College, Cambridge, and £250 from the Madras University. The communication from Madras was received later than the one from Trinity. The sum of £500 appeared to him far too much. According to his Gospel of Wealth, any money which comes to one's share beyond the actual needs was only to be received as Trust Fund for social good. It should be used only for the benefit of those that were in need. It is this Gospel of Wealth that made him write a letter to the University of Madras on the receipt of its offer. Ramanujan was then living in Putney, a nursing home in London. He was too weak to write the letter himself. This letter was therefore in Hardy's hand. Here is the text of that letter:

> 2 *Colinette Road, Putney, SW* 15
> 11 *January* 1919

To
   The Registrar,
   University of Madras.
Sir,
   I beg to acknowledge the receipt of your letter of 9th December 1918, and gratefully accept the very generous help which the University offers me.
   I feel however, that, after my return to India, which I expect

103

to happen as soon as arrangements can be made, the total amount of money to which I shall be entitled will be much more than I shall require. I should hope that, after my expenses in England have been paid, £50 a year will be paid to my parents and that the surplus, after my necessary expenses are met, should be used for some educational purpose, such in particular as the reduction of school-fees for poor boys and orphans and provision of books in schools. No doubt it will be possible to make an arrangement about this after my return.

I feel very sorry that, as I have not been well, I have not been able to do so much mathematics during the last two years as before. I hope that I shall soon be able to do more and will certainly do my best to deserve the help that has been given me.

I beg to remain, Sir,
Your most obedient servant
(S RAMANUJAN)

Srinivasa Ramanujan

Entrance Door of Ramanujan's House in Triplicane
(Photograph by C Seshachalam)

Inscription on Memorial Tablet over the Entrance
(Photograph by C Seshachalam)

Mrs Ramanujan
(*Photo*: *C Seshachalam*)

Ramanujan Commemoration Stamp on "First-day" Cover
(*Photograph by C Seshachalam*)

COLINETTE HOUSE,
2, COLINETTE ROAD,
PUTNEY, S.W.15.

8645

To the Registrar of the University of
Madras

11 January 1919

Sir / beg to acknowledge the receipt
of your letter of 9 Dec 1918,
and gratefully accept the very
generous help which the University
offers me.

I feel, however, that, after
my return to India, which I
expect to happen as soon as
arrangements can be made, the
total amount of money to be

[opposite

I shall be entitled will be made 3/
more than I shall require. I should
hope that, after my expenses in
England have been paid, £50 a
year will be paid to my parents,
and that the surplus, after my
necessary expenses are met, should
be used for some educational
purpose, such in particular
as the reduction of school fees
for poor boys and orphans and
provision of books in schools. No
doubt it will be possible to
make an arrangement about
this after my return.

I feel very sorry that, as I
have not been well, I have
not been able to do so much

[over

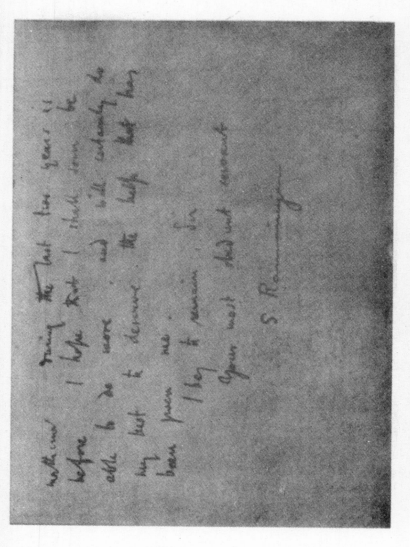

Copy of Ramanujan's letter to the Registrar, University of Madras, from a nursing home in Putney, London. The letter was written in G H Hardy's hand. (For text see pp. 103-104)

(With Acknowledgements to the University of Madras)

RAMANUJAN, THE MATHEMATICIAN

P1 Early Indications

The first occasion when Ramanujan was known to have shown unusual interest in Mathematics was when he was 12 years old. He is then said to have asked a friend studying in a higher class of the Town High School in Kumbakonam about the "highest truth" in Mathematics. The theorem of Pythagoras and the problem of Stocks and Shares were said to have been mentioned to him as the "highest truth"! He was then studying in the third form. One day, his teacher in Arithmetic said something like the following: "If three fruits are divided among three persons, each would get one. Even if 1,000 fruits are divided among 1,000 persons, each would get one." From this, he generalized that any number divided by itself was unity. This made Ramanujan jump up and ask, "Is zero divided by zero also unity? If no fruits are divided among nobody, will each get one?" This was perhaps the first indication of his unusual insight into the behaviour of numbers. In the same year, he is said to have worked out the properties of the Arithmetical, Geometrical, and Harmonic Progressions. When he was 13, Ramanujan borrowed a copy of Loney's *Trigonometry* from a student in the Degree Class of the Government College, Kumbakonam, and solved all the problems in the book. In the next year, Ramanujan derived Euler's Expansions of the Circular Functions in infinite series, all by himself.

P11 HIGHER FLIGHTS

As already stated in Sec CO, it was in 1903, when he was 15, that Carr's *Synopsis* stimulated him. Between 1904 and 1911, his famous "Frayed Notebook" got filled up, chapter by chapter. He began with Magic Squares. The major problems engaging his thought in those years related to Integers, including Distribution of Prime Numbers and of Highly Composite Numbers, Partition of Numbers, Integers expressible as sums of squares, cubes, etc, Elliptic Integrals, Hypergeometric Series, and Definite Integrals of various kinds. The main results he had obtained before 1913 on these topics, were communicated to Hardy in his letter dated 16 January 1913 and 27 February 1913.

P2 Fractional Differentiation

In 1914, a reception was given, perhaps jointly to Neville and Ramanujan, on the eve of their leaving for England. At that time, Ramanujan mentioned the idea of Fractional Differentiation. The Differential Co-efficient of x^n is nx^{n-1}. Ramanujan expressed n as

$$\frac{1.2.3 \ldots n}{1.2.3 \ldots (n-1)}$$

This is read as

$$\frac{\text{Factorial } (n)}{\text{Factorial } (n-1)}$$

We can also differentiate x^n twice and get as a result
$$n\,(n-1)\,x^{n-2}$$
The co-efficient here may be written as
$$\frac{\text{Factorial } (n)}{\text{Factorial } (n-2)}$$

Differentiation can be repeated thrice, four times, etc.

In this approach, we can have only "integral differentiation". That is, differentiation can be repeated only an integral number of times. For, "Factorial (n)" is meaningful only when n is a positive integer. But in the meeting, Ramanujan suggested as follows:

Factorial $n = n$ times Factorial $(n-1)$, where n is a positive integer.

We have the function Gamma (n) defined by a well-known definite integral.

When n is an integer, Gamma $(n+1) = n$ times Gamma (n),
Thus when n is an integer,

Gamma $(n+1)$ is identical with Factorial n.
But the function Gamma $(n+1)$ has meaning also when n is a fraction. Therefore if,

D = Differential co-efficient;

D^2 = Second differential co-efficient; and in general

106

$D''' = m$th differential co-efficient,
We can write,

$$Dx^n = \frac{\text{Gamma } (n+1)}{\text{Gamma } n}\ x^{n-1}$$

$$D^2x^n = \frac{\text{Gamma } (n+1)}{\text{Gamma } (n-1)}\ x^{n-2}$$

$$D'''x^n = \frac{\text{Gamma } (n+1)}{\text{Gamma } (n-m+1)}\ x^{n-m}$$

Now put $m = \frac{1}{2}$

$$\text{Then, } D^{\frac{1}{2}} x^n = \frac{\text{Gamma } (n+1)}{\text{Gamma } (n+\frac{1}{2})}\ x^{n-\frac{1}{2}}$$

This fact enables us to have fractional differentiation. We the students attending the function were wondering at this insight and generalization by Ramanujan. This trivial result is stated here in such details to enable non-mathematicians to get an idea of the way in which Ramanujan's mind penetrated into regions of Mathematics. In this connection, the following extract from Ramanujan's letter of 16 January 1913 to Hardy may be of interest:

"Just as in elementary mathematics, you give a meaning to a^n when n is negative and fractional, to conform to the law which holds when n is a positive integer, similarly the whole of my investigations proceed on giving a meaning to Eulerian Second Integral for all values of n. My friends who have gone through the regular course of University education tell me that the integral of
$$x^{n-1}\ e^{-x} = \text{Gamma } (n)$$
is true only when n is positive. They say that this integral relation is not true when n is negative. Supposing this is true only for positive values of n and also supposing the definition n Gamma $(n) =$ Gamma $(n+1)$ to be universally true, I have given meanings to these integrals and under these conditions I state that the integral is true for all values of n — negative and fractional. My whole investigations are based upon this and I have been developing this to a remarkable extent so much so that the local mathematicians are not able to understand me in my higher flights." What is

107

given below on the value of his work on these topics is all culled from the writings of G H Hardy.

P3 Prime Number Problem

The prime number problem, still not fully solved, concerns the number of prime numbers less than n. This is one of the most fascinating problems of pursuit in the whole field of mathematics. P V Seshu Ayyar had shown Ramanujan, the copy of Hardy's tract on *Orders of infinity* in the library of the Presidency College, Madras. But what this tract led Ramanujan to, is described in his own letter to Hardy as follows:

P31 RAMANUJAN WAS NEVER DIM

Ramanujan wrote further on in the letter of 16 January 1913, "Very recently I came across a tract published by you styled *Orders of infinity* in page 36 of which I find a statement that no definite expression has been as yet found for the number of prime numbers less than any given number. I have found an expression which very nearly approximates to the real result, the error being negligible....I have found a function which exactly represents the number of prime numbers less than x — "Exactly" in the sense that the difference between the functions and the actual number of primes is generally zero or some small finite value even when x becomes infinite. I have got the function in the form of infinite series and I have expressed it in two ways." In Hardy's book on Ramanujan (1940), Chapter 2 is devoted to this theorem. He points out the flaws in Ramanujan's arguments and finally concludes, "I do not think that it is worthwhile to probe more closely into the details of Ramanujan's reasoning. There is one reproach at any rate that cannot be made against it. Unsound as it is, it is not "dim"; Ramanujan was never dim. It contains a very interesting idea, and one which, when properly pruned, fits into its place in the theory."

P32 FINE FORMAL IDEAS

"Whatever one may think of Ramanujan's argument, one will agree that his formal ideas were fine, and one is sure to wonder whether they were all his own. In particular, did he really discover the Riemann's series himself?"

108

P33 NO ACCESS TO WHITTAKER

My own opinion is that he did. It seems worthwhile to consider for a moment what books, of importance to him and accessible in Madras, Ramanujan may have consulted. Hardy refers to my having sent him a copy of the catalogue of the Madras University Library published in 1914. In that catalogue were listed Whittaker's *Modern Analysis*, Bromwich's *Infinite Series* and Mathew's *Theory of Numbers*. Hardy wonders how these three books could have escaped his attention. Hardy could not have known that the Madras University Library was not accessible to Ramanujan till after he became a research scholar of that University in May 1913. Moreover that library was just then being formed and had not been thrown open to the public and brought into use till 1914. Hardy himself finds internal evidence in Ramanujan's letter for his not having had access to the University Library. For, Hardy proceeds, "In the first place, Ramanujan *cannot* have seen Whittaker, since he did not know Cauchy's Theorem."

P34 NO ACCESS TO BROMWICH

"The evidence about Bromwich is not so conclusive, but I cannot believe that Ramanujan had seen the book. He was keenly interested in divergent series, about which he had a 'theory' of his own. Bromwich has a long and very interesting chapter on the subject, which would have fascinated Ramanujan; but Ramanujan never showed any knowledge of Cessaro or Borel Summability, or of any of the standard works. It seems plain indeed from passages in his letters that he had no idea that any scientific theory of divergent series existed."

P35 ACCESS TO GREENHILL

"I do not think, then, that Ramanujan had seen either Whittaker or Bromwich; but it seems plain that he had read some book on elliptic functions, and I agree with Littlewood in thinking that it was probably Greenhill's. He never refers to books, but he never referred to any of the standard theorems of the subject though he thought them his own. He claims to have extended the theory in different directions, as he had done, but not to have invented elliptic integrals, theta functions, or modular equations. All these things he treats as parts of common knowledge. His

109

own knowledge was remarkable both for its extent and for its limitations, and both the extent and the limitations fit excellently with the hypothesis that it was based on Greenhill's stimulating but eccentric book."

P36 NO ACCESS TO MATHEWS

"The theorems about prime numbers, on the other hand, Ramanujan claimed quite definitely as his own (though of course he recognised his mistakes later, and learnt the outlines of the established theory). This, to any one who knew Ramanujan well, is conclusive; but the hypothesis of Ramanujan's complete independence is also the only one which seems to me to fit the facts."

"In the first place, if Ramanujan had ever seen Mathews, how could he have been as ignorant as he was of the classical theory of quadratic forms, which Mathews discusses elaborately and which fills nearly half of his book? The series for the class-number, in particular, would have fascinated Ramanujan, and it is certain that he would have studied them intensively."

"But the most conclusive evidence is Mathew's chapter on primes itself. This, whatever its defects, contains a fairly adequate description of Riemann's memoir."

"My conclusion is, therefore, that all this work of Ramanujan, with its flashes of inspiration and its crude mistakes, was an individual and unassisted achievement. There is no other hypothesis which seems to me to be tenable, or to make any sort of mathematical or psychological sense."

P37 ASTONISHING PERFORMANCE

"It may be stated here that Ramanujan had access only to copies of Hardy's Tract already mentioned and perhaps Greenhill's *Elliptic functions* in the Presidency College Library, which Seshu Ayyar might have lent to him." [My conjecture is that S Narayana Ayyar should have brought Greenhill to his attention.]

"In conclusion I may say this. No one before Riemann, so far as I know, had written down $R(x)$, and if Ramanujan found the series himself, it may seem a very astonishing performance. Even Gauss stopped at li x."

P4 Highly Composite Numbers

A highly composite number is, in a sense, the very opposite of a prime number. A prime number has only two divisors — itself and unity. A highly composite number has more divisors than any preceding number. For example, the first few highly composite numbers are 2, 4, 6, 12, 24, 36, 48, 60, 120.

Collected Papers pages 87 and 88 contain a table of the first 103 Ramanujan's highly composite numbers. The last of these is a huge number. It is 6 746 328 388 800.

Ramanujan studied the structure, the distribution, and the special forms of highly composite numbers. In teaching the concept of highly composite numbers, I have found that the student's attention can be drawn, with the aid of the graphs of the number of divisors of n, to the parallelism between the erratic distribution of prime numbers and that of highly composite numbers. On Ramanujan's work of highly composite numbers, Hardy writes that it "represents work, perhaps, in a back-water of mathematics, and is somewhat over-loaded with detail; but the elementary analysis of "highly composite" numbers... is most remarkable and shows very clearly Ramanujan's extraordinary mastery over the algebra of inequalities!"

P5 Partition of Integers

A partition of integer n is a division of n into any number of positive integral parts. For example,

$$4 = 3 + 1 = 2 + 2 = 2 + 1 + 1 = 1 + 1 + 1 + 1$$

We say that 4 has five partitions. In symbols we write it as

$$p(4) = 5$$

The number of partitions of n is denoted by $p(n)$.
This is called a Partition Function.

According to Hardy, "Very little is known about the arithmetical properties of $p(n)$ when n is odd or even. Ramanujan was the first, and up to now, the only mathematician to discover any such properties; and his theorems were discovered, in the first instance, by observation." With the aid of the theory of elliptic functions, he proved some of the congruent properties of the Partition Function. Some of these conjectures were found to be true and some of them found to be contradictory by actual calculation. Hardy speaks of two formulae known as "Rogers-Ramanujan

111

Identities." About them he writes:

"The formulae have a very curious histo:y. They were found first in 1894 by Rogers, a mathematician of great talent but comparatively little reputation, now remembered mainly from Ramanujan's re-discovery of his work. Rogers was a fine analyst, whose gifts were, on a smaller scale, not unlike Ramanujan's; but no one paid much attention to anything he did, and the particular paper in which he proved the formulae was quite neglected.

"Ramanujan re-discovered the formulae sometime before 1913. He had then no proof (and knew that he had none), and none of the mathematicians to whom I communicated the formulae could find one. They are therefore stated without proof in the second volume of MacMahon's *Combinatory analysis*.

"The mystery was solved, trebly, in 1917. In that year Ramanujan, looking through old volumes of the *Proceedings of the London Mathematical Society*, came accidentally across Rogers' paper. I can remember very well his surprise and the admiration which he expressed for Rogers' work. A correspondence followed in the course of which Rogers was led to a considerable simplification of his original proof. About the same time I Schur, who was then cut off from England by the war, re-discovered the identities again. Schur published two proofs, one of which is "combinatorial" and quite unlike any other proof known. There are now seven published proofs, — the four referred to already, the two simpler proofs of Schur, and the one found later by Rogers and Ramanujan and published in the *Collected Papers* of Ramanujan."

For the first time, Hardy and Ramanujan jointly examined in 1917 the question of how large the number of partitions of n is when n itself is large. They gave the answer in the form of an asymptotic series and also estimated the error involved in taking a definite number of terms only. As an example, we may reproduce here the number of partitions of 14031. It is

$$p\,(14031) = 92\ 85303\ 04759\ 09931\ 69434\ 85156\ 67127$$
$$75089\ 29160\ 56358\ 46500\ 54568\ 28164$$
$$58081\ 50403\ 46756\ 75123\ 95895\ 59113$$
$$47418\ 88383\ 22063\ 43272\ 91599\ 91345$$
$$00745$$

Commenting on this, Hardy observes that the prediction by Ramanujan that the number of partitions of 14031 would be divisible by 114 has, turned out to be true.

P6   A Number as a Sum of Squares

The problem of representation of an integer n as the sum of a given number k of integral squares is one of the most celebrated in the theory of numbers. The most famous of these theorems is the Fermat Theorem that a prime number of the form $4m + 1$ is the sum of two squares. This is yet to be proved. There are three considerable papers of Ramanujan on this subject. They occur as papers 18, 20, and 21 of his *Collected Papers*.

P61   ANECDOTE 8: EVERY INTEGER A FRIEND OF RAMANUJAN

According to Hardy, Ramanujan could remember the idiosyncracies of numbers in an almost uncanny way. He further states that according to Littlewood, every integer was one of the personal friends of Ramanujan.   Ramanujan was in a sanatorium in Putney. Hardy went to see him.

HARDY: I came in the taxi-cab 1729.   It is rather a dull number. I hope it is not an unfavourable omen.

RAMANUJAN: No, it is a very interesting number.

HARDY: How?

RAMANUJAN: It is the smallest number expressible as the sum of two cubes in two ways. $(1729 = 1^3 + 12^3 = 9^3 + 10^3)$.

HARDY: What is the smallest number expressible as the sum of fourth powers?

RAMANUJAN: (After thinking for a moment) I see no obvious example.   I think that the number must be very large.

After narrating this anecdote, Hardy refers to a number given by Euler

$$N = 158^4 + 59^4 = 134^4 + 134^4$$

P7   Other Subjects

P71   HYPERGEOMETRIC SERIES

The most considerable and complete work done by Ramanujan in India before he left for England in 1914 appears to be the one on the Hypergeometric series. As stated in Sec K2, Hardy published in 1923 Ramanujan's results on this subject, with proofs and annotations in the *Proceedings* of the Cambridge Philosophical Society. This stimulated a flood of papers by Bailey, Watson, Whipple, and others. These have been brought together in W N Bailey's *Generalised Hypergeometric series* (1935), (Cambridge Tracts on Mathematics and Physics). The key formula on the subject had been found by Ramanujan about 1910 or 1911. But no proof had been given by him in his Notebook. Hardy states that in this formula Ramanujan had been anticipated by Dougall, though Ramanujan had not known it. Hardy, therefore, calls this formula Dougall-Ramanujan formula. Hardy further says that he himself found this formula in Ramanujan's Notebook only after his death. The Notebook of Ramanujan contains a mass of elegant summations. A few of these had been already found and proved by Heine (1878), Dickson (1891), and Morley (1902). But Ramanujan had had no access to these. What he could gather from books available to him was nothing more than what was contained in Chrystal's *Algebra*.

P72 ALGEBRA

According to Hardy, Ramanujan's main work in Algebra was concerned with both Hypergeometric Series and Continued Fractions, "These subjects suited him exactly, and here he was unquestionably one of the greatest masters." Ramanujan's "Frayed Notebook" contains some of his masterpieces in Continued Fractions. On this subject also, he had no other start than what he could find in Chrystal's *Algebra*. According to Hardy, "It is perhaps in his work in these fields that Ramanujan shows at his very best.... It was his insight into algebraical formulae, transformation of infinite series, and so forth that was most amazing. On this side, most certainly I have never met his equal, and I can compare him only with Euler and Jacobi.... It is possible that the great days of formulae are finished, and that Ramanujan ought to have been born 100 years ago; but he was by far the greatest formalist of his time. There have been a good many more important, and I suppose one must say greater, mathematicians than Ramanujan during the last fifty years, but not one who could stand up to him on his own ground. Playing the game of which he knew the rules, he could give any mathematician in the world, fifteen."

P73 DEFINITE INTEGRALS

Before Ramanujan went to England, he was mostly engaging himself in working out general formulae in Definite Integrals. During the nine months of his research studentship of the Madras University, he produced quite a number of results in this field. These are all included in his three quarterly reports to the University. According to Hardy, "He had no real proofs of any of the formulae." Hardy goes on to explain the connotation of the term "real proof."

"It is reasonable to say that we now know, roughly, the conditions for the truth of most analytical formulae. They do not intrude merely on account of the weakness of our analysis, but are genuine limitations corresponding broadly to the facts. Our theorems will not cover all cases in which the formula is true, and it may be interesting and profitable to do what we can to extend them; but the conditions under which we have proved them would become insufficient if widened in any really drastic way.

"A mathematician may have stated a formula and advanced reasons for its truth which are inadequate as they stand, in which case he cannot be said to have "proved" it. But it often happens that his method, when re-stated and developed by a modern analyst, leads to a proof valid under "natural" conditions, and in that case we may fairly say that he has "really" proved the theorem. Thus Euler "really" proved large parts of the classical analysis, and there are a great many theorems which Ramanujan had "really" proved; but he had not "really" proved any of the formulae I have quoted. It was impossible that he should have done so because the "natural" conditions involve ideas of which he knew nothing in 1914, and which he had hardly absorbed before his dealth. He had also, as Littlewood says, "no-clear-cut conception of proof. If a significant piece of reasoning occurred somewhere and the total mixture of evidence and intuition gave him certainty, he looked no further." In this case any "real" proof was inevitably beyond his grasp, and the "significant pieces of reasoning which are indicated in the notebooks and reports,

115

though we shall find them curious and interesting, are quite inadequate" as proof.

P74 ELLIPTIC FUNCTIONS

Ramanujan had been handling elliptic functions profusely. About this subject, Hardy says, "Ramanujan never professed to have made any major advance in the general theory of Elliptic Functions." According to Littlewood, "Ramanujan somehow acquired an effectively complete knowledge of the formal side of the Theory of Elliptic Function." According to Hardy, "Ramanujan shows at his very best in the parts of the Theory of Elliptic Functions" allied to the Theory of Partitions.

The last and perhaps the only letter written to Hardy by Ramanujan after his return to India is dated 12 January 1920. In this letter he speaks of "Mock-Theta Function." He gives in that letter a list of such functions up to order seven (*See* Sec B5).

P8 Overall Evaluation

General overall evaluation of Ramanujan, the mathematician, has been given in the following words by J J Thompson, Master of Trinity College, Cambridge, in his book *Recollections and Reflections* (1935).

There were in his notebooks statements without proofs of a large number of theorems. These have been worked over by several eminent mathematicians, who have succeeded in proving the correctness of a good many of them, and thereby greatly strengthened the verdict of Professor Hardy that in his own field he was unrivalled in his day. His method, however, was not the normal one in which the theorem arises out of the proof; no proof, no theorem. It is possible, however, to imagine other ways of proving theorems. Suppose, for example, a mathematician dreamt that he had discovered a new theorem, if he remembered it when he awoke he might test it by seeing if it gave the right result in a great number of special cases. This is the method of "trial and error"; the great difficulty in this method is to get something to try; there are an infinite number of things which might be tried, and unless we had something to guide us the chance of choosing the right one would be infinitesimal. It need not require a dream to act as a guide. One who, like Ramanujan, had made a long

116

and intense study of a particular branch of mathematics might almost unconsciously have been led to recognize certain features, such as the absence or presence of certain arrangements of the symbols in the theorems known to be true, and would instinctively reject a theorem in which these did not occur. He would, by long experience, have acquired an instinct by which he could distinguish between theorems which were possible and those which were impossible. Then, if he had the imagination to think of a theorem which would satisfy the test and the industry and power of calculation required to verify it, he might arrive at theorems which he could not prove. There are several instances of mathematical theorems which are believed to be true but have never been proved. Perhaps the most famous is the formula given by Gauss for the number of prime numbers (i. e. numbers which, like 2, 3, 5, 7, 11... are not divisible by any other numbers), which are less than a given number N; Gauss's formula was tested for integral values of N up to a thousand millions, and was found to give the right result. This was universally accepted as overwhelming evidence of its truth, though no formal proof had been discovered. Professor Littlewood has shown, however, that it must ultimately fail when N is greater than a certain number. This number, however, is so prodigious that it would be beyond the power of human effort to count by trial the number of primes. Thus Gauss's rule may console itself by thinking that though it may lapse from rectitude it never does so when it can be found out.

WORKS BY AND ON RAMANUJAN

Q1 Works by Ramanujan

1 to 3   Notebooks of Ramanujan;

4 to 6   The Quarterly Reports furnished by him to the University of Madras in 1913-1914; and

7   His *Collected Papers* published in 1927 by the Cambridge University Press.

Q2 Published copies of the Notebooks

The Notebooks have been brought out in photostat form by the Tata Institute of Fundamental Research, Bombay, as stated in Sec L8.

Q3 Quarterly Reports

It is stated that G N Watson had copied out the three quarterly reports to the University of Madras in a copy of the Notebooks of Ramanujan sent to Hardy in 1926. The University of Madras may publish these three quarterly reports.

Q4 Collected Papers

The *Collected Papers* contain all the five papers published in the *Journal* of the Indian Mathematical Society before he left India, all the thirty-two papers published after he reached Cambridge, and all the fifty-eight questions published in the *Journal* of the Indian Mathematical Society between the years 1911 and 1919.

Q5 Questions and Solutions

Of the fifty-eight questions, Ramanujan himself furnished solutions to the following seven questions:

284, 289, 294, 295, 327, 507, and 584

Solutions to the following eleven questions do not appear to have been given by anybody.

352, 387, 526, 662, 699, 722, 738, 755, 784, 1076 and 1094

The remaining forty questions have been solved by various mathematicians, mostly Indian and mostly in the *Journal* of the Indian Mathematical Society.

It was a happy idea of the Founder of the Society, V Ramaswamy Ayyar, to have instituted the pages "Questions and Solutions". I remember with what avidity some of us young men used to look forward to these pages from issue to issue.

Q6 Books on Ramanujan

The following two books have been published on Ramanujan:

1 Hardy (G H). Lectures on the mathematical work of Ramanujan, reproduced by Marshall Hall, 1936. (Institute for Advance Studies, Princeton).

2 Hardy (G H). Ramanujan: Twelve lectures on subjects suggested by his life and work, 1940. (Cambridge University Press).

The second of the above books gives the fuller version of the first book and in addition, the substance of several other lectures delivered by Hardy between 1936 and 1940, at the Harvard Ter-Centenary Conference of Arts and Sciences, and to a number of Universities and Societies in U K and in U S A.

Q7 Papers on Ramanujan

The second book of Hardy gives a bibliography of 110 papers on Ramanujan. Of these, six are on his life. Some give also an evaluation of his work. Of these six, four were published in the *Journal* of the Indian Mathematical Society, one in the *Mathematical Gazette* and in *Nature*, and one in the *Proceedings* of the London Mathematical Society and in the *Proceedings* of the Royal Society (*A*). The remaining 104 papers are original papers on Ramanujan. The following table gives a list of the authors of these papers.

SN	Author					Number of papers
1	Bailey (W N)	.	.	.	.	4
2	Chowla (S)	.	.	.	.	1
3	Copson (E T)	.	.	.	.	1
4	Darling (H B C)	.	.	.	.	3
5	Erdos (P)	.	.	.	.	1
6	Estermann (T)	.	.	.	.	1
7	Goodspeed (F M)	.	.	.	.	1
8	Gupta (H)	.	.	.	.	3
9	Hardy (G H)	.	.	.	.	11
10	Hodgkinson (J)	.	.	.	.	1
11	Ingham (A E)	.	.	.	.	2
12	Krecmar (W)	.	.	.	.	1
13	Lehmer (D H)	.	.	.	.	5
14	MacMahon (P A)	.	.	.	.	2
15	Mordell (L J)	.	.	.	.	6
16	Narayana Ayyar (S)	.	.	.	.	2
17	Page (A)	.	.	.	.	1
18	Phillips (E G)	.	.	.	.	2
19	Pillai (S S)	.	.	.	.	1
20	Preece (C T)	.	.	.	.	6
21	Rademacher (H)	.	.	.	.	5
22	Rankin (R A)	.	.	.	.	3
23	Rogers (L J)	.	.	.	.	3
24	Stanley (G K)	.	.	.	.	1
25	Szego (G)	.	.	.	.	1
26	Turan (P)	.	.	.	.	2
27	Watson (G N)	.	.	.	.	25
28	Whipple (F J W)	.	.	.	.	3
29	Wilson (B M)	.	.	.	.	1
30	Wilton (J R)	.	.	.	.	2
31	Zuckermann (H S)	.	.	.	.	3
						104

Four of the thirty-one authors are Indians. They are: 1 Chowla (S), 2 Gupta (H), 3 Narayana Ayyar (S), and 4 Pillai (S S). They have together contributed 7 of the 104 papers.

Q71   Distribution by periodical and by Country

Country	Periodical	N in Periodical	N in Country
1 India	Indian Mathematical Society		
	Journal	4	
	Indian Academy of Sciences		
	Proceedings	1	5
		—	
2 Great Britain	Cambridge Philosophical Society		
	Proceedings	14	
	Transactions	1	
	Edinburgh Mathematical Society		
	Proceedings	1	
	London Mathematical Society		
	Journal	43	
	Proceedings	17	
	Messenger of mathematics	2	
	Oxford quarterly journal of mathematics	5	
	Quarterly journal of mathematics	2	85
		—	
3 U S A	American journal of mathematics	1	
	American Mathematical Society		
	Bulletin	2	
	Transactions	2	
	Annals of mathematics	2	
	Duke mathematical journal	1	
	National Academy of Sciences		
	Proceedings	1	9
		—	
4 Germany	Journal fur Mathematik.	1	
	Mathematische Zeitschrift	1	2
		—	
5 U S S R	Academy of Sciences		
	Bulletin	1	1
		—	
6 Sweden	Acta mathematica	1	1
		—	
7 Poland	Acta arithmetica	1	1
		—	
			104

Q72   DISTRIBUTION BY YEAR

Year	N of Papers	Cumulative N of Papers
1913	2	2
1915	1	3
1917	2	5
1919	3	8
1920	3	11
1921	4	15
1922	1	16
1923	2	18
1924	2	20
1926	3 ⎤	23
1927	2 ⎥	25
1928	8 ⎥	33
1929	10 ⎬ 36	43
1930	4 ⎥	47
1931	9 ⎦	56
1932	5	61
1933	5	66
1934	2	68
1935	3	71
1936	7 ⎤	78
1937	12 ⎥	90
1938	5 ⎬ 31	95
1939	7 ⎦	102
1940	2	104

The six-year period 1926 to 1931 saw the publication of 36 papers, that is, nearly one-third of the total. This is largely due to the fact that a copy of the Notebooks of Ramanujan reached Hardy in 1926. Again, the four-year period 1936-1939 saw the publication of 31 papers, that is, a little less than one-third. This is largely due to lectures on Ramanujan by Hardy, before a number of universities and societies in the USA during the period.

Q73   CHECK-UP

It is desirable to check up if the above-mentioned bibliography is complete up to 1940, and if it is not, to make it complete.

UP-TO-DATE BIBLIOGRAPHY

R1 Target Year: 1967

There is no doubt that many more papers would have come out on Ramanujan since 1940. I expect that many papers would have been stimulated during the last five years by the photostat copies of Ramanujan's Notebooks published by the Tata Institute of Fundamental Research. A suitable epoch to bring the bibliography up-to-date is 22 December 1967; for, it marks the completion of the 80th year of Ramanujan.

R2 Agency

What should be the agency for such a bibliography. One can think of three agencies — the University of Madras, to which he belonged, the Indian Mathematical Society, to which also he belonged in a sense, and the University of Cambridge which accepted him in 1914. It occurs to me that it is in the fitness of things that this bibliography should be prepared by an Indian agency. For, the conditions in India today have changed completely from what they were in the days of Ramanujan. India is now an independent country. It now has a good number of able mathematicians. It has now also a number of competent documentalists. It would be unworthy of Independent India to allow this work to be done by an outside agency. Moreover, practically all the worthwhile periodicals containing mathematical contributions are now available in India. Indeed, the Madras University Library itself has been very rich in its collection of mathematical books and periodicals, since 1926.

R3 A Suggestion

The University of Madras has a Department of Library Science. It has also the Ramanujan Institute housed in it. It has further a Department of Mathematics. These three institutions belonging to the University of Madras can jointly take up the publication of the Ramanujan bibliography. The two mathematical institu-

123

tions can make a thorough search in periodicals and in books for all references — direct as well as indirect — to Ramanujan. They may also furnish to each contribution a brief indicative abstract. In this work, the two institutions may also invoke the help of the Indian Mathematical Society. The work can be organized on a co-operative basis. The host documents to be searched may be divided among the various mathematicians. A Standard may be furnished by the Department of Library Science for recording each item picked up, on separate slips of standard size. Of course, this standard will be in conformity with the specification of the Indian Standards Institution, in respect of the diverse sections of an entry such as the heading, the title section, the notes section, the host section, and the abstract section. In the case of a book, abstract will give place to descriptive annotation. The Department of Library Science may then take over all the collected slips, classify them, arrange them with feature headings, and provide the necessary alphabetical, chronological, and other kinds of index entries. The classification will require considerable sharpening of the existing schedules. In fact, it will require depth classification of mathematics, more formidable than what has been hitherto designed. In this work, the Department of Library Science of the University of Madras will do well to work in collaboration with the Documentation Research and Training Centre in Bangalore, which is an All India Institution.

R4 Publication

The completion of the search and the organization of the resulting slips may perhaps take about four years, if the work is to be done entirely on honorary basis. However, the University of Madras can reduce this period in two ways. It can entrust the search for materials and abstracting work to one of the research students in the Department of Mathematics. Similarly, it can appoint a research student in the Department of Library Science to carry out the further work of organizing the slips and preparing the press copy. It would be useful to make this research student to work part of the time in the Documentation Research and Training Centre in Bangalore, in order to get the necessary depth classification designed. In this arrangement, the search can be completed in one year. Similarly, the organization of the slips and the preparation of the press copy can be completed in another one year. These two periods, of one year each, can also be telescoped to some extent. The printing and the publishing work may take about a year. Thus the exhaustive bibliography can be published, by about 1970.

R5 Future

If the present generation of Indian mathematicians and documentalists do this work for the epoch marked by the 80th year of Ramanujan, there is every chance that this work will be continued by the next generation and that an up-to-date bibliography will be brought out with much less effort and with much more thoroughness, as a Centenary Bibliography brought upto the end of 1987. This will make the thoroughness of the centenary bibliography greater, because it is possible that it would be difficult to find some of the earlier materials 25 years later. It is a matter of gratification that our country has begun to remember our great men of recent years and celebrate their centenaries and other epochs. This is a sign of the renaissance of our country today. This kind of work is necessary to enable our future generations know of the richness of their heritage. Such a knowledge will stimulate them and accelerate their own further march forwards.

BIO-DATA

22 Dec 1887	Born at Erode in the house of his maternal grandfather, who was Amin in the District Munsiff's Court at Erode. Father — Srinivasa Ayyangar Mother — Komalammal Grandmother — Rangammal In his bodily build Ramanujan was very much like his mother and grandmother. Wife — Janaki, born in 1901.
1888 to 1914	Lived in his hometown Kumbakonam in Sarangapani Street, where his parents lived, his father being a clerk in a cloth merchant's shop.
1899	Precocity noticed first.
1903	Reads Carr's *Synopsis*.
Dec 1903	Passes Matriculation Examination of the University of Madras, in First Class.
1904	Studies in the First Year of the F A Class in the Government College, Kumbakonam, with the Subramanyan Scholarship.
1905	Discontinues study and spends some months in Andhra Pradesh.
1906	Resumes study in Pachaiappa College. Lives with grandmother in a house in a lane near the Fruit Bazaar, George Town. Falls ill. Does not appear for the examination.
1907 to 1908	Lives mostly in Kumbakonam.
1907	Appears for F A Examination but fails.
1909	Marries Janaki. Lives in Madras, but falls ill and goes back to Kumbakonam. Was operated for kidney trouble.
1910 to April 1912	Lives in Summer House, Sami Pillai Street, Triplicane, Madras.

1910 to 1911	Meets V Ramaswamy Ayyar, Founder of the Indian Mathematical Society and R Ramachandra Rao, President of the Indian Mathematical Society, shows his Notebook to them and asks for a clerical post.
1911	Supported for a few months by R Ramachandra Rao.
From 1 March 1912 to 30 April 1913	Clerk in the Accounts Department of Madras Port Trust. First lives in Saiva Muthiah Mudali Street, George Town and later in Hanumantharayan Koil Street, Triplicane, Madras.
1 May 1913 to 16 March 1914	Gets Research Scholarship. First lives in Hanumantharayan Koil Street and later in Thope Venkatachala Mudali Street, Triplicane Madras.
16 Jan 1913	First letter to G H Hardy.
8 Feb 1913 to January 1914	G H Hardy's endeavour to have Ramanujan in Cambridge.
Early March 1914	Gets Overseas Scholarship from the University of Madras.
17 March 1914	Sails from Madras to London.
1916	Gets Honorary B A Degree of the University of Cambridge.
1918 to 1920	Attack of Tuberculosis.
1918	Elected Fellow of the Royal Society and awarded Fellowship of the Trinity College, Cambridge. Madras University Scholarship extended to five more years.
11 Jan 1919	Last letter written to the University of Madras.
27 March 1919	Arrives in Bombay.
2 April 1919	Arrives in Madras.
12 Jan 1920	Last letter to G H Hardy.
26 April 1920	Dies in Chetpet, Madras

INDEX

by M A GOPINATH

Note 1: The index number is the whole number of the chapter, or section of the occurrence of the item indexed.
Note 2: The following contractions are used:

irt = in relation to
qirt = question in relation to
rby = referred by

rin = referred in
rirt = referred in relation to

Academic setting in
  Cambridge E7
  Madras E43
Achievement B7
Activity period
  1 C
  2 G
Adinarayanan L5
Agency for bibliography R2
Alagappa Chettiar K31
Algebra P72
All India Radio A5
Ananda Rao ML
Ancient lore MJ7
Anecdote
  1 C1
  2 D2
  3 D3
  4 D4
  5 D6
  6 K33
  7 L3
  8 P61
Archives D8
Asia Publishing House A6
Astrology *rirt*
  Ramanujan
    irt
      His death B24
      His greatness MC4
      His reputation B24
    rby Janaki MT5
Aurobindo N31
Avasara N5
Avatara N5

B

Bailey P71
Beach Railway Station MJ91
Bed MN2
Berry MN1
Bhabha L8

Bharadwaja N27
Bibliography R
Biography A2
  Genesis of A6
Birth
  Astrology on B22
  Date of B7
Blanket MN2
Board of studies E34
Books Q6
Borel P34
Brahma N27
Bromwich P34

C

Cambridge
  Call of E62
  ‾Friends ML/MN
  Mandate from E82
Cap incident MF3
Carr's *Synopsis*
  as trigger book C0
  Introduction to C1
Centenary bibliography R5
Cessaro P34
Chandrasekhara Ayyar J6
Chandrasekharan (K) *rirt*
  Notebook
    Editing of L72
    Publication of L8
Chandrasekharan (S) *rirt*
  Portrait K1
  Ramanujan Prize K2
Chengalvarayan MD
Chetpet N22
Chinese gentleman L72
Christ N5
Chronology of notebooks L6
Chrystal's *Algebra* P71
Climate of England F7
Cock and bull story B3

Index

Collected papers
Proposal for G4
irt Its fulfilment G5
rirt
  Achievement B72
  Biography *irt*
    Psycheo-genetic force B6
    Trans-rationality B3
College
  career C3
  friends MD/MK
Common Wealth Univ Cong L71
Communication after death B5
Conjecture *rirt*
  Intuition N42
  Its depth ML7
Conspectus A7
Croydon anecdote F3

D

Daily practices 'N7
  rby Janaki MT6
Date of
  birth B7
  death J6
  voyage
    abroad F15
    home J5
Death J6
  Astrology on B24
  rby
    Janaki MT6
    Narasimha Ayyangar MG7
Definite integral P73
Delight N25
Deshmukh MM
Devaraja Mudaliar ME
Dickson P71
Dougall P71
Drashta C6
Dream *irt*
  Elliptic integrals MQ1
  Precocity B22
  rby Gopalachary MP42
  Voyage B23
Dress problem F6
Driver's pride MS1
DRTC
  rirt
    Bibliography R3
    Depth classification R4
    Students A4
Dualism N5
Duraiswamy Pillai MF61

E

Economic sufficiency D

Elliptic
  function P74
  integral *rby* Berry MN1
  integral *rby*
    Narayana Ayyar D6
    Rajagopalan MQ1
Emotional
  qualities N23
  strain J1
  warmth N25
Employment D
English scale of values J11
Esdaile L4
Euler P1
  rby Hardy P72
European dress MB2
Exhibition of Trinity College J3
Exhibition on Ramanujan K5
Eyes N1
  rby
    Mahalanobis MN6
    Radhakrishnan MJ2

F

False story C4
Family
  ,allowance F13
  life MT1
Fellowship of
  Royal Society B71
    irt Date H1
  Trinity College H2
Finance Ministry K34
Flair N411
Food habit *rby*
  Ananda Rao ML2
  Deshmukh MM4
  Janaki MT3/4
  Neville (Mrs) F5
  Ranganathan J2
  Spring F14
Fractional differentiation P2
Frayed Notebook
  see Notebook 1
Friendliness
  see Sociability
Frog—See
  Human frog
  Sea frog
  Well frog
Funny anecdotes MJ4

G

Ganapathy Subbier MC2
Ganesa Sastry MP3
Gangasnanam MF4

Gauss P8
Genius
  Explanation of B6
  rin Anecdote 4 D4
Getting into Bed MN2
God, Zero, Infinity
  in Metaphysics N51
  rby Gopalachari MP1
    Mahalanobis MN5
Gopala Ayyar MJ92
Gopalachary MP
Gospel of Wealth N8
Government
  College C3
  of India K34
Governor's tribute MF7
Govindarajan MC
Grandmother
  rby
    Janaki MT3
    Radhakrishnan MJ3
  rirt Namagiri B22
Greenhill P35
Griffith E1
Gwyer L7

H

Hanuman N1
Hanumantha Rao *rirt*
  Board of Studies E3
  Notebook 1 MP6
Hardy
  and Ramanujan *irt*
    Partition of numbers P5
    Reciprocal learning G1
  irt
    Cambridge and Ramanujan *irt*
      Attempt 1 E63
      Attempt 2 E64
    Grooming Ramanujan G2
    Modesty of Ramanujan N27
    Notebook 1 L3
    Number 1729 P61
    Orders of infinity P3
    Ramanujan *irt*
      Algebra P72
      Bromwich P34
      Definite integrals P73
      Elliptic function P74
      Fractional differentiation P2
      Greenhill P35
      His intuition N44
      His letter
        First E61
        Last P74
      Inequalities P4
      Integer P61

Mathews P36
Mock-theta function P74
Overseas scholarship E6
Partition function P5
Research scholarship MG3
Whittaker P33
Ranganathan *irt*
  Collected papers L3
  Notebook 1 L3
Vijayaraghavan E62
qirt Ramanujan P
Ramanujan and Littlewood G2
rirt
  Carr's *Synopsis* C1
  Collected papers G4
  Fractional differentiation P2
  Illness J4
  Photograph K1
  Ramanujan *irt*
    Riemann P37
    Rogers P5
  Ramanujan's
    first letter E61
    last letter P74
Religious practice of Ramanujan N5
Walker E2
Health of Ramanujan J
Heine P71
Higher mathematics P11
Hill E1
Hindu MT4
Honours to Ramanujan H
Horoscope of
  Janaki B24
  Ramanujan B24
Human frog MF61
Humility E61
Humour N22
Hyde Park anecdote F3
Hyper-geometric series P71

I

ICS probationer B3
Idol worshipper L4
Ill at ease F2
Illness *rby*
  Deshmukh MM6
  Gopalachari MP42
  Hardy J4
  Janaki MT6
  Narasimha Ayyangar MG7
  Radhakrishnan (R) MJ8
Indian
  food J2
  Mathematical Society *irt*
    Felicitations H4
    Bibliography R2

Index

Notebook L8
National archives D8
Industry N3
  rby Thompson P8
  rin Anecdote 4 D4
Infinity, God, and Zero—
  See God, Zero, Infinity
Innocence MD1
Institute of Fundamental
  Research K32
Integer P5
Intellect N4
Intellectual
  flair N411
  life J3
International Cong Math K32
Intuition N4
  rby Berry MN1

J

Jacobi P72
  rby Ananda Rao ML7
Janaki *irt*
  Her horoscope B24
  Monthly allowance L1
  Reminiscences MT
  Wit and humour N22
Jayasinghe A6
Jesus Christ N5

K

Kartru-tantra N4
Kasturiranga Ayyangar MT4
Kodumudi N22
Kosambi K2
Krishnan L8
Krishna Rao ML1
Krishna Sastriar MF3
Krishnaswamy Ayyangar B5
Krishnaswami Ayyar MF
Kundalini-sakti MF62

L

Lakshmanaswamy Mudaliar *rirt*
  Notebooks and their
    Editing L72
    Publishing L71
Last days of Ramanujan J6
  ·*rby*
    Janaki MT6
    Narasimha Ayyangar MG7
    Radhakrishnan MJ93
Legal hurdle E4
Library Science Department R3
Life in Cambridge

Preparation for F12
rby
  Ananda Rao ML2
  Deshmukh MM2
  Janaki MT4
  Mahalanobis MN
  Narasimha Ayyangar MG6
Limiting point of Intellect N41
  Littlehailes *irt*
  Ramanujan
    Outfit of F12
    Professorship H5
Littlewood *irt*
  Conjecture N44
  Elliptic function P74
  Ramanujan G2
London Mathematical Society P5
Loney's *Dynamics of a particle* MN3
Loney's *Trigonometry*
  rin Anecdote 1 C1
  rirt Ramanujan's ability P1

M

MacMahon P5
MacPhail B6
Madras Port Trust *irt*
  Anecdote 5 D6
  Employment D5
  Future D7
Madras Christian College MJ91
Magic squares P11
Mahabharata MJ7
Mahalanobis MN
Mahasaraswathi N31
Man, The N
Mathematical
  ability P1
  meteor J7
Mathematician, The P
Mathematics
  Department R3
  Higher P11
  Teaching of G11
  Under cot MP7
Mathews *rirt*
  Pillai K33
  Ramanujan P36
Matriculation examination C2
Maya rby
  Mahalanobis MN5
  Venkatarama Ayyar MS1
Memento MT7
Memorial K
Mental
  absorption *rby*
    Govindarajan MC1
    Janaki MT3

132

Krishnaswami Ayyar MF4
Srinivasaraghavacharya MK2
qualities N1
Meta-Physics *irt*
  Equations MR2
  Zero to infinity N51
    rby Mahalanobis MN5
Meteoric career A1
Middlemast D5
Missing Notebook, *See* Notebook 1
Mock-theta function *irt*
  Hardy P74
  Ouja Board B5
Modesty *rby*
  Devaraja Mudaliar ME2
  Hardy N27
Morley P71
Mother N31
Mother of Ramanujan
  irt
    Astrology B4
    Her dream B23
  rby
    Ananda Rao ML4
    Janaki MT3
Mysticism MR2

N

Nagoya J5
Namagiri *irt*
  Birth B22
  Dream on child's death MP42
  Manifestation of God N5
  Precocity B22
Namakkal—*See* Namagiri
Narasimha Ayyangar MG
Narasimha, Lord, *rby*
  Radhakrishnan MJ6
  Rajagopalan (TK) MQ2
Narayana Ayyangar F11
Narayana Ayyar *irt*
  Anecdote 5 D6
  Board of studies E3
  Employment of Ramanujan D5
  Interest in Ramanujan D7
  Voyage of Ramanujan B23
  Walker E2
Narayanaswami MJ8
Narayanaswamy Ayyar B24
Nevasa F15
Neville (Mr) *irt*
  Hardy's attempt 2 E64
  Outfit of Ramanujan F12
  Reception P2
Neville (Mrs) F
Nir-Guna-Brahman N51
  rby Mahalanobis MN5

Notebook
  Chronology of L6
  Copying of L6
  irt Its publication Q2
Notebook 1 *irt*
  Its recovery L3
  Its search L1
  Its value L4
  Ramachandra Rao D3
    irt Copying L6
  Ramanujan's inner light C6
  Ramaswamy Ayyar D2
  Ranganathan L3
  Ross MJ91
  Singaravelu Mudaliar MJ91
Notebook 3 B5
  rby
    Radhakrishnan MJ91
    Thompson P8
Number
  7 MH1
  1729 P61
  Friend's Society ML
Numberumal Chetty J6

O

Obstinate Patient MJ8
Occult
  experience MP4
  phenomenon
    Belief in MP4
    See also Trans-rational
    phenomenon
Occultism N6
Open-set N41
Ouja Board B5
Outfit F12
Output of research G3
Overseas scholarship
  Efforts for E5
  rby Narasimha Ayyangar MG4

P

Pachiappa College
  irt College career C3
  rby Srinivasa Raghavacharya MK1
Packing paper MK3
Papers Q7
Paranjpye H4
Partition function P5
Patanjali MF62
Patanjali Sastry MH
Patience of Ramanujan MF4
Patient MJ8
Patrachariar MP1
Pedlar's pill MP2
Pantland MF7

Index

Personal qualities N2
Personality MN6
Philosophy N5
  rby
    Gopalachari MP1
    Mahalanobis MN5
    Radhakrishnan MJ7
Physical
  appearance *rby*
    Radhakrishnan MJ2
    Devaraja Mudaliar ME4
  strain J2
Physiology MF6
Piccadilly J2
Pillai *rirt*
  His death K32
  Notebook K33
  Professorship K32
Pittendrig B1
Polish mathematician B2
Pongal MT4
Portrait K1
  rby Janaki MT7
  rby Krishnaswami Ayyar   MF8
Postage stamp K4
Poverty *rby*
  Krishnaswamy Ayyar MF3
  Radhakrishnan MJ3
Precocity B22
Pre-cognition MR3
Presidency College Math Assoc MF7
Press note F16
Prime number problem P3
Private Secretary's letter E83
Proceedings, Lond Math Soc P5
Professorship H5
Psycheo-genetic force B6
Psychology B1
Publication of
  bibliography R4
  Notebooks L8
Purva-janma-vasana B6
Pythagoras theorem P1

Q

Quarterly reports Q3

R

Radhakrishna Ayyar *Algebra* MD1
Radhakrishnan (R) *irt*
  Religious lore ME5
  ·Reminiscences MJ
Radhakrishnan (S) L71
Raghunathan MB
Rajagopalan (C T) K34
Rajagopalan (T K) MQ

Ramachandra Rao
  irt
    Chronology of Notebooks L6
    Interview D3
      rby Radhakrishnan MJ92
    Notebook 1 D3
    Research scholarship E1
    Seshu Ayyar D3
    Support of Ramanujan D4
  rirt
    Biography B3
    Collected papers G4
    Tuft F12
    Voyage F11
Ramakrishna Paramahamsa N5
Ramana Maharshi N5
  Eyes of N1
Ramanujachari *rby*
  Devaraja Mudaliar ME2
  Krishnaswami Ayyar MF
  Radhakrishnan MJ5
Ramanujan
  and Hardy—*See* Hardy
  archives D8
  as astrologer MT5
  as mystic MR2
  as patient MJ8
  as tutor MC3
  Bio-data of S
  Biography of—*See* Biography
  College career of C3
  Exhibition K5
  Eyes of—*See* Eyes
  groomed
    for Cambridge F1
    in mathematics G2
  Hall K6
  ill at ease F2
  Institute K3
    irt Bibliography R3
Ramanujan *irt*
  Algebra P2
  Applied Mathematics MN3
  Astrology—*See* Astrology
  Borel P34
  Bromwich P34
  Carr's *Synopsis* CO
  Chetpet N22
  Cessaro P34
  Chrystal's *Algebra* P71
  Collected papers—
    See Collected papers
  Collecting packing paper MK3
  Continued fraction MN4
  Cooking *rby*
    Deshmukh MM3
    Janaki MT4
    Mahalanobis MN4

134

Definite integral   P73
Dickson   P71
Delight   N25
Digestive system   MF63
Dougall   P71
Dualism   N5
Economic sufficiency   D
Elliptic
    functions   P74
    integral—*See* Elliptic integral
Emotional strain   J1
Euler   P72
Felicitations   H4
Fellowship—*See* Fellowship
Fractional differentiation   P2
Frog—*See* Frog
God, Zero, Infinity—*See* God
Government College   C3
Greenhill   P35
Hardy—*See* Hardy
Heine   P71
Highly evolved souls   MP41
His
    belief in
        God   MR2
        occultism   B4
    cap   MF3
    coaching   MG1
    communication after death   B5
    conjecture—*See* Conjecture
    daily practices—
        See Daily Practices
    death—*See* Death
    dream—*See* Dream
    dress   F6
    emotional qualities   N23
    employment   D5
    first letter to Hardy   E61
    food—*See* Food habit
    friendliness   N26
    genius   D4
    grandmother—*See* Grandmother
    greatness   MC4
    health   J
    horoscope   B24
    humility   E61
    illness—*See* Illness
    industry—*See* Industry
    initiation   MP2
    innocence   MD1
    interview with
        Ramachandra Rao   D3
        Ramaswamy Ayyar   D2
        Seshu Ayyar   D2
    interest *rby*
        Ananda Rao   ML5
        Gopalachary   MP3
    intellectual life   J3

intuition—*See* Intuition
knowledge of math   G1
last letter to
    Hardy   P74
    Univ Madras   N8
last days—*See*   last days
letter to
    Hardy
        First   E61
        Last   P74
    Univ Madras   N8
life in Cambridge—
    See Life in Cambridge
marks in F A   C5
matriculation exam   C2
mental absorption—
    See Mental absorption
metaphysics—*See* Metaphysics
modesty—*See* Modesty
notebook—*See* Notebook
occult experience   N6
outfit   F12
output of research   G3
patience   MF4
personality   MN6
philosophy   N5
physical appearance—
    See Physical appearance
portrait—*See* Portrait
poverty—*See* Poverty
precocity   B22
precognition   MR3
proficiency   MB1
religious practices—
    See Religious practices
reputation   B24
simplicity   MM5
sociability—*See* Sociability
spiritual experience   MQ1
talent   MH2
temper   MP5
temperament   N21
rby Mahalanobis   MN6
tuft—*See* Tuft
turban   F3
unassuming quality   ML6
voyage—*See* Voyage
warmth   N25
wit—*See* Wit
works   Q1
wordly factors   DO
Honours   H
Hyper-geometric series   P71
Integer   P61
Jacobi—*See* Jacobi
Janaki—*See* Janaki
Karma   MP3
Kasturiranga Ayyangar   MT4

Ramanujan *irt*
Krishna Rao   ML1
Littlewood—*See* Littlewood
London Mathematical Society   P5
Mahabharata   MJ7
Mathematical meteor—
  See Mathematical meteor
Mathews   P36
Mock-theta function—
  See Mock-theta function
Morley   P71
Moksha   MP3
Namagiri—*See* Namagiri
Narasimha—*See* Narasimha
Notebook—*See* Notebook
Number—*See* Number
Occult—*See* Occult
Overseas Scholarship—
  See Overseas Scholarship
Pachiappa College—
  See Pachiappa College
Partition function   P5
Philosophy   MJ7
Polish mathematician   B2
Professorship   H5
Ramachandra Rao—
  See Ramachandra Rao
Ramaswamy Ayyar—
  See Ramaswamy Ayyar
Ramayana   MJ7
Religious practice—
  See Religious Practice
Research scholarship—
  See Research Scholarship
Riemann   P37
Rogers   P5
Ross—*See* Ross
Royal Society—*See* Royal Society
Sandow   D4
School—*See* School
Seshu Ayyar—*See* Seshu Ayyar
Social practices   N7
Subramaniam scholarship   C3
Super-activity period—
  See Super-activity period
Support by Ramachandra Rao   D4
Tanjavur   N22
Time-table   MC2
Town High School—
  See Town High School
Trinity College—
  See Trinity College
Vishnu-sahasra-nama   B5
Ramanujan
  memento   MT7
  memorial   K
  on look out for employment   D1
  postage stamp   K4

Prize   K2
returns home   J5
the Man   N
the Mathematician   P
Ramanujan (Mrs)—*See* Janaki
Ramaswamy Ayyar
  irt   Seshu Ayyar   D2
  rin   Anecdote 2   D2
  rirt
    Employment   D5
    Felicitation   H4
    Voyage   F11
Ramayana   MJ7
Ranganathan   *irt*
  Chinese gentleman   L72
  Chronology of Notebooks   L6
  Esdaile   L4
  Examination statistics   C5
  Hardy   *irt*
    Madras Univ Lib Catalogue   P33
    Notebook 1   L3
  His turban   F3
  Nevelle (Mr)   F2
  Nevelle (Mrs)   F⁻
  Pillai   K3
  Vijayaraghavan   E62
Reciprocal learning   G1
Religious
  experience   N5
  lore   ME5
  practice   MR1
    and social practice   N7
Reminiscences   M
Research scholarship   E41
  irt   Gospel of Wealth   N8
Riemann   P37
Ritual   J1
Rogers-Ramanujan identity   P5
Ross
  irt   Report   B2
  qirt   Polish mathematician   B2
  rby
    Narasimha Ayyangar   MG3
    Ranganathan   L3
  rirt   Notebook   MJ91
Royal Society
  irt   Fellowship   B71
    irt   Date   H1

S

Sadhu Sundar Singh   N5
Sa-Guna-Brahman   N51
Sandow   D4
Sankara   *rirt*
  Intuition   N43
  Religious practice   N5
Sankhya   MP8

Santhanam D8
Sanatoria at Wells J4
Sarva-Guna-Asraya N51
Satyapriyarayar MP41
Scale of values J11
Scholarship
    Overseas E83
    Research E41
    Sharing of MF1
School
    Career C2
    friends MB/MC
    time-table MC2
Sea-frog MF61
Seer in mathematics C6
Serpent theory in Yoga MF62
Seshu Ayyar
    irt
        Examination statistics C5
        Ramachandra Rao D3
        Ramanujan *irt*
            Biography B3
            Discovery B1
            Employment D4
            Greenhill P35
            Orders of infinity P3
    rirt
        Collected papers G4
        Ramanujan's proficiency MB1
        Ramaswamy Ayyar D2
        Voyage of Ramanujan F11
Seventy-fifth birthday A5
Sivasankaranarayana Pillai—*See* Pillai
Shyness N21
    rby Mahalanobis MN6
Simplicity
    rby Deshmukh N27
Singaravelu Mudaliar *irt*
    His appreciation on
        Ramanujan *rby*
            Devaraja Mudaliar ME3
            Krishnaswamy Ayyar MF5
            Radhakrishnan MJ6
        Notebook MJ6
Sita N1
Sociability N26
    rby
        Ananda Rao ML3
        Chengalvarayan MD1
        Mahalanobis MN6
        Radhakrishnan MJ4
Social practices N7
South Indian communities J11
Space-time-junction point MP42
Spiritual
    help MP1
    qualities N1
Spring *irt*

Anecdote 5 D6
Future of Ramanujan D7
His letter to Private Secretary E8
Ramanujan's food F14
Srinivasa Ayyangar (K) K1
Srinivasa Ayyangar (S) J6
Srinivasan (G A) L5
Srinivasan (K S) D4
Srinivasan (R) MR
Srinivasa Raghavacharya MK
SS Nevasa F15
SS Nagoya J5
Stocks and shares P1
Strand Magazine MN4
Subbanarayanan (N) B23
Subramanian scholarship C3
Sundaram Ayyar E4
Superactivity period
    1 C
    2 G
Suryanarayana Sastry K1
Swaminatha Ayyar MP6
Symbolic metaphysics N51
Syndicate minute E41

                T

Taming Ramanujan N44
Tanjavur N22
Taxi-cab 1729 P61
Temper MP5
Thompson P8
Time-space-junction point MP42
Town High School
    irt Time-table MC2
    rin Anecdote 1 C1
    rirt Mathematical ability P1
Transcendental order of genius E81
Trans-rational
    information B21
    phenomenon
        Allergy to B3
        See also Occult phenomenon
Trigger book CO
Trinity College
    irt
        Exhibition J3
        Fellowship H2
    rby
        Ananda Rao ML2
        Deshmukh MM1
        Janaki MT4
        Mahalanobis MN1
        Narasimha Ayyangar MG6
Trivikrama Rao E43
True story C5
Tuft
    rby
        Neville (Mrs) F4

Index

Radhakrishnan MJ93
Turban problem F3

U

Unassuming quality ML6
Undigestible digestion MF63
University
  of Cambridge *rirt*
    Bibliography R2
    Overseas scholarship E7
  of Madras *rirt*
    Academic setting 1 E43
    Bibliography R2
    Exhibition K5
    Notebook L4
      irt Its recovery L1
    Overseas scholarship E7
    Professorship H5
    Ramanujan Institute K34
    Ramanujan Prize K2
    Research
      allowance H3
      scholarship *irt*
        Legal hurdle E4
        Sanction E41
        Walker E2
  setting E
  system C4

V

Vaidyanathaswamy L5
Valmiki *rirt*
  Eyes of Rama N1
  Humility of Rama E61
Vastu-tantra N4
Venkatarama Ayyar MS
Venkatraman D8
Vijayaraghavan

irt Ramanujan C4
rirt
  Hardy E62
  His death K34
  Oxford G11
  Ramanujan Professorship K31
Vishnu-sahasra-nama B5
Vivekananda MF61

W

Wali Mohammad L71
Walker *rirt*
  Board of Studies E3
  University of Madras E3
Watson L5
Wealth N8
Whittaker P33
Well-frog MF61
Wilson G5
Winners of Ramanujan Prize K2
Wit N22
  rby
    Janaki MT6
    Krishnaswami Ayyar MF6
    Venkatrama Ayyar MS1
Works by and on Ramanujan Q
Wordly factors DO

Y

Yates MF1
Yoga MF62

Z

Zero
  irt Metaphysics N51
  rby
    Gopalachari MP1
    Mahalanobis MN5

Time spent on compiling the index:

SN	Stage	Hours
1	First reading of the text (140 typed pages)	6
2	Preparation of slips. Individual entries (819)	44
3	Verification with text	12
4	Alphabetisation	6
5	Consolidation of entries (360)	12
6	Final verification of the press copy	10
	Total	**90**